Spatial Analysis in Soil Dynamics and Earthquake Engineering

Proceedings of sessions held in
conjunction with Geo-Logan '97
sponsored by The Geo-Institute of
the American Society of Civil Engineers

Utah State University
Logan, Utah
July 16-19, 1997

Edited by J. David Frost

Geotechnical Special Publication No. 67

Published by the
ASCE *American Society of Civil Engineers*
345 East 47th Street
New York, New York 10017-2398

Abstract:

This Geotechnical Special Publication contains the papers presented by the invited speakers in the sessions on "Spatial Analysis in Soil Dynamics and Earthquake Engineering" held during the First National Conference of the ASCE Geo-Institute. The conference was hosted by Utah State University in Logan, Utah from July 16 to 19, 1997. Topics presented in the sessions included seismic hazard analysis in research and practice, regional versus site specific analysis methods, spatial data quality evaluation, spatial ground motion amplification, spatial liquefaction analysis, zonation of seismic slope instability, multi-hazard spatial analysis and spatial assessment of seismic induced lifeline damage. This document provides the engineer, geologist, seismologist, planner or risk manager with a comprehensive and up-to-date perspective of the opinions of selected researchers and practitioners from academia, government and industry on the use of spatial analysis techniques for evaluating earthquake hazards and consequences.

Library of Congress Cataloging-In-Publication Data

Spatial analysis in soil dynamics and earthquake engineering : proceedings of sessions held in conjunction with Geo-Logan '97 : sponsored by the Geo-Institute of the American Society of Civil Engineers : Utah State University, Logan, Utah, July 16-19, 1997 / edited by J. David Frost.
p.   cm. -- (Geotechnical special publication ; no. 67)
ISBN 0-7844-0258-2
1. Soil dynamics--Statistical methods. 2. Spatial analysis (Statistics) 3. Earthquake engineering--Statistical methods. I. Frost, J. David II. American Society of Civil Engineers. Geo-Institute. III. Geo-Logan '97 (1997 : Logan, Utah) IV. Series.

TA711.S653 1997   97-20894
624.1'762'01519535--dc21   CIP

GEOTECHNICAL SPECIAL PUBLICATIONS

1) TERZAGHI LECTURES
2) GEOTECHNICAL ASPECTS OF STIFF AND HARD CLAYS
3) LANDSLIDE DAMS: PROCESSES, RISK, AND MITIGATION
4) TIEBACKS FOR BULKHEADS
5) SETTLEMENT OF SHALLOW FOUNDATION ON COHESIONLESS SOILS: DESIGN AND PERFORMANCE
6) USE OF IN SITU TESTS IN GEOTECHNICAL ENGINEERING
7) TIMBER BULKHEADS
8) FOUNDATIONS FOR TRANSMISSION LINE TOWERS
9) FOUNDATIONS AND EXCAVATIONS IN DECOMPOSED ROCK OF THE PIEDMONT PROVINCE
10) ENGINEERING ASPECTS OF SOIL EROSION, DISPERSIVE CLAYS AND LOESS
11) DYNAMIC RESPONSE OF PILE FOUNDATIONS— EXPERIMENT, ANALYSIS AND OBSERVATION
12) SOIL IMPROVEMENT - A TEN YEAR UPDATE
13) GEOTECHNICAL PRACTICE FOR SOLID WASTE DISPOSAL '87
14) GEOTECHNICAL ASPECTS OF KARST TERRAINS
15) MEASURED PERFORMANCE SHALLOW FOUNDATIONS
16) SPECIAL TOPICS IN FOUNDATIONS
17) SOIL PROPERTIES EVALUATION FROM CENTRIFUGAL MODELS
18) GEOSYNTHETICS FOR SOIL IMPROVEMENT
19) MINE INDUCED SUBSIDENCE: EFFECTS ON ENGINEERED STRUCTURES
20) EARTHQUAKE ENGINEERING & SOIL DYNAMICS (II)
21) HYDRAULIC FILL STRUCTURES
22) FOUNDATION ENGINEERING
23) PREDICTED AND OBSERVED AXIAL BEHAVIOR OF PILES
24) RESILIENT MODULI OF SOILS: LABORATORY CONDITIONS
25) DESIGN AND PERFORMANCE OF EARTH RETAINING STRUCTURES
26) WASTE CONTAINMENT SYSTEMS: CONSTRUCTION, REGULATION, AND PERFORMANCE
27) GEOTECHNICAL ENGINEERING CONGRESS
28) DETECTION OF AND CONSTRUCTION AT THE SOIL/ROCK INTERFACE
29) RECENT ADVANCES IN INSTRUMENTATION, DATA ACQUISITION AND TESTING IN SOIL DYNAMICS
30) GROUTING, SOIL IMPROVEMENT AND GEOSYNTHETICS
31) STABILITY AND PERFORMANCE OF SLOPES AND EMBANKMENTS II (A 25-YEAR PERSPECTIVE)
32) EMBANKMENT DAMS-JAMES L. SHERARD CONTRIBUTIONS
33) EXCAVATION AND SUPPORT FOR THE URBAN INFRASTRUCTURE
34) PILES UNDER DYNAMIC LOADS
35) GEOTECHNICAL PRACTICE IN DAM REHABILITATION
36) FLY ASH FOR SOIL IMPROVEMENT
37) ADVANCES IN SITE CHARACTERIZATION: DATA ACQUISITION, DATA MANAGEMENT AND DATA INTERPRETATION
38) DESIGN AND PERFORMANCE OF DEEP FOUNDATIONS: PILES AND PIERS IN SOIL AND SOFT ROCK

FOREWORD

The National Science Foundation sponsored a workshop in January, 1993 on "Geographic Information Systems and Their Application in Geotechnical Earthquake Engineering". Participants included researchers and practitioners from academia, government and industry who were brought together to exchange information and ideas relating to GIS technology and its application in geotechnical earthquake engineering and to provide guidelines for others using or considering the use of the technology in the future. The proceedings of that two-day workshop were published by ASCE and contained the papers submitted by the workshop attendees, the written reports from the working groups and other general information pertaining to the workshop program. While there have been a number of papers and isolated sessions at a variety of conventions, conferences and symposia over the past few years, it was considered to be very timely for the ASCE Geotechnical Engineering Division and in particular, the Soil Dynamics Committee to organize a more focused series of sessions at a major conference so that a number of key issues relating to the use of spatial analysis technologies (of which Geographic Information System [GIS] technology is one of the key ones) could be presented and discussed in an open forum.

As a result, sessions have been organized as part of the First Geo-Institute Conference and have been designed to provide ample opportunity for formal presentation of developments in the area of spatial hazard analysis while at the same time ensuring that session attendees can raise additional issues for discussion. Presenters have again been selected from government, industry and academia thus varied opinions and experiences can be expected.

The initial papers in the sessions include a comprehensive overview of the topic (Kiremidjian), followed by a presentation which focuses on identifying issues related to implementation of the technology in practice (Rogers). Two subsequent presentations deal with issues resulting from the scale used in the analysis (Mabey) and data quality evaluation (Luna). The next three presentations focus on spatial analysis within the context of specific earthquake hazards including ground motion amplification (Borcherdt), liquefaction (Frost) and earthquake induced landslides (Ho). These are followed by papers which look at techniques for assessing the combined hazard from multiple threats (King) and discussion of techniques for damage estimation and lifeline infrastructure evaluation within a GIS framework (O'Rourke). Each of the presenters in the sessions was specifically requested not to present the results of their own work only but to discuss the topic from a broader perspective. Obviously, they were free to illustrate a particular point with an example from their own work, but the intention was to ensure a broader perspective of developments in a particular topic is provided.

A number of people's activities made the sessions possible and thus also contributed to the preparation of these proceedings. The ASCE Geotechnical Division Soil Dynamics Committee under the Chairmanship of Panos Dakoulas first proposed the idea of organizing these sessions. The subsequent support of the organizing committee of the Geo-Logan '97 conference was vital including, in particular, the efforts of Leslie Youd and Priscilla Nelson in supporting the idea of a proceedings with color figures. The willingness of the invited speakers to make the time to prepare the manuscripts and to personally present their ideas and findings at the conference has played an important role in making the sessions and the proceedings a reality. Special recognition and thanks are due to Stacy Lewis for designing the cover of the proceedings and Dan Carroll for help with final editing. Finally, the patience, perseverance and assistance of Shiela Menaker, Manager of Book Production at ASCE, was, as always, most important.

Each of the papers included in the proceedings has been accepted for publication by the proceedings editor. All papers are eligible for discussion in the Journal of Geotechnical and Geoenvironmental Engineering and are eligible for ASCE awards.

J. David Frost
Editor

CONTENTS

Spatial Analysis in Geotechnical Earthquake Engineering

Anne S. Kiremidjian[1], M. ASCE

Abstract

Geographic information systems (GIS) have demonstrated to be a valuable tool in geotechnical earthquake engineering. They provide the means for storing, manipulating, analyzing and graphically representing soil data and analysis results. Utilization of these systems in geotechnical earthquake engineering has become particularly important because of the spatial character of the data and models in this field. GIS are now extensively used for ground motion hazard computations, landslide and liquefaction analysis over large regions and more localized ground deformation assessments. This paper summarizes the various ways in which GIS has been utilized in geotechnical earthquake engineering, brings out issues that need to be considered and discusses future directions for the field.

Introduction

Inherent in all geotechnical problems is the need for spatial data storage, integration, analysis and visualization. Geographic information systems (GIS) provide the facilities for performing these functions and as such have become an integral tool of geotechnical earthquake engineering problem solutions. Such systems can be utilized at different scales depending on the type of problem and the level of accuracy required in the analysis. Currently, 2D - GIS are extensively used and 3D GIS systems are becoming ever more popular as the storage and computational capacity of computers are increasing.

The three main problems of concern in geotechnical earthquake engineering include estimation of potential ground shaking, liquefaction and landslide susceptibility over areas. These analysis are performed either for scenario

---

[1] Professor and Director, The John A. Blume Earthquake Engineering Center, Department of Civil Engineering, Stanford University, Stanford, CA 94305-4020

earthquakes (deterministic analysis) or using probabilistic hazard analysis which considers the effect from all possible future events. Following susceptibility evaluations, more detailed analysis can be performed on a local scale leading to ground deformation and affected area computations.

In this paper, the general framework of GIS-based analysis for geotechnical earthquake hazards is presented first. Then, example applications are used to illustrate some of the concepts discussed in the paper. Key issues are briefly summarized and some thoughts on future directions are presented with the concluding remarks.

<u>Framework for GIS-based Geotechnical Earthquake Hazard Analysis</u>

Figure 1 shows the general flow of GIS-based hazard estimation that can be used for any of the three types of analyses regardless of whether probabilistic or deterministic earthquake is used. From this figure, the main components of a GIS-based approach include (a) data compilation, (b) data integration, manipulation cleansing, (c) analysis and (d) output.

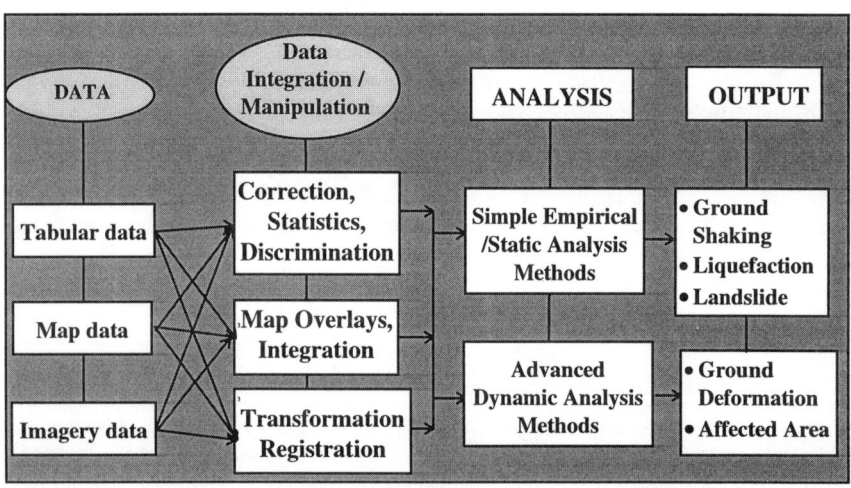

Figure 1. Schematic representation of modeling within GIS.

The most difficult, time consuming and frequently the most costly part of this process is the development of the databases. Three types of data are used in GIS based evaluation. These include tabular, map and imagery data. Tabular data is most frequently stored in a database management system (DBMS) to enable easy access and manipulation of the information from the different modules of the system. Data needed for geotechnical investigations, however, varies with depth below ground surface requiring representation in three dimensional space. For investigations that cover small area and for which considerable number of bore hole data are available, 3D GIS provides the tools for storing the detailed information with depth. For studies over large geographic areas, however, such detailed information is not available and even if it were available, the problems of storage, efficient access and manipulation are significant. Average values of important parameters are used for such studies. For example, average mass density, shear wave velocity, and depth to ground water for the top 30 m of soil may be stored for centroids of polygons with uniform surface geologic units or for grids of specified size (e.g., $0.1°$ x $0.1°$ grids).

The geometric and topological aspects of geographic information, however, cannot be stored in traditional database management systems. The utility of GIS is the greatest for these data bases. With the release of satellite imagery data in the commercial domain and the decrease in cost of aerial photography, these information sources are now being used extensively to supplement map data. GIS provides the tools for manipulating, overlaying and integrating the information from the various map, imagery and conventional DBMS. Combined with advanced analytical tools, these comprehensive packages have enabled engineers to perform studies on larger scales than ever before and to provide the profession and the public with more accurate and detailed information.

Ground Motion Amplification Within GIS

At a particular site, various advanced analytical models are available to conduct in-depth studies provided local soil parameter information is available. Effects of local topography, such as basins and ridges, can be included when the scale of the application is relatively small. For such applications GIS can be a useful data storage and display tool. On a more regional scale, however, detailed soil data are not available and smoothed or average parameters at grids or polygons are used.

Currently, there is an effort to generate a soil parameter database from borehole information in California (e.g., Vucetic and Doroudian, 1995). This database contains 850 digitized borehole data. In a recent study by Vucetic et al., (in Kiremidjian et al, 1997), similar borehole data were used to evaluate earthquake ground motion amplification/deamplification using the one-dimensional non-linear model in *DESRAMOD2* (modified by Doroudian et al. 1996 from *D-MOD* by Matasovic and Vucetic, 1993) to predict surface ground motions in a small area within the city of Palo Alto. Figure 2 shows the area within Palo Alto and the 73 borehole locations that were used in that study. Figure 3 shows the cross section of boring logs and bedrock profile. In building this database (stored in *Techabse*[TM]) a comprehensive data structure was developed to include parameters that contained information on stratification, ground water table, soil classification based on Plasticity Index and silt content, moisture content, unit weight, SPT blow count, and shear wave velocity. Since different data sources were used to compile the

3

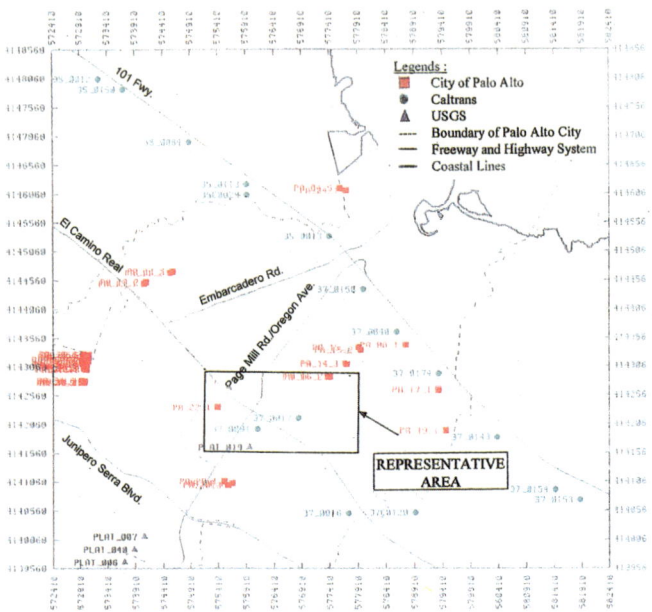

Figure 2. Map of boring log locations in the study region within Palo Alto, California (Kiremidjian et al., 1997)

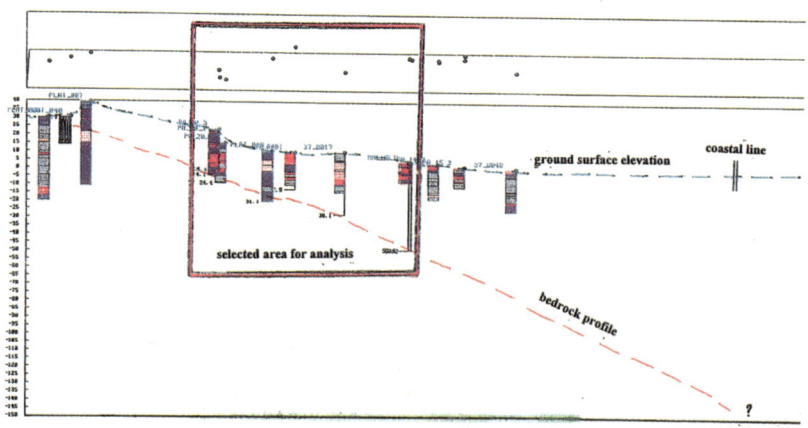

Figure 3. Cross section of boring logs and bedrock profile (Doroudian and Vucetic, in Kiremidjian et al., 1997)

information in the database, considerable effort was undertaken to bring the data to a unified classification and to infer missing information from available data.

The information from the database was used in DESRAMOD-2 to ground motion amplification/deamplification studies. Figure 4 shows the map of maximum ground surface acceleration for the study area. A generic ground motion time history, modified from one of the local records from the 1989 Loma Prieta earthquake was used for this purpose. In addition, the borelog data was combined with a map of thickness of alluvium and bay mud for Palo Alto within *Techabse*[TM] to produce the depth to bedrock needed for ground motion amplification estimation. Often, ground motion amplification/deamplification maps can be used in microzoning. Such a map is shown in Figure 5. Other types of maps that can be produced in the same way include maximum shear stress, predominant period of soil columns or from ground surface acceleration response spectra, maximum normalized pore water pressure, and maximum shear stress and shear strain. Moreover, an ensemble of time histories corresponding to a constant risk bedrock spectra can be generated and used to produce similar maps corresponding to a specified risk level.

On the larger more regional scales, however, it is not feasible to perform such detailed analysis. Most frequently, empirical relationships are used to modify bedrock motions or surface ground motions are predicted using simple soil classification scheme. For example, the Boore et al.(1993) ground motion attenuation relationship was used to develop bedrock ground motions for scenario earthquakes as well as for a specified probability of exceedince (Kiremidjian et al., 1997). These were generated with the use of ARC/INFO[TM]. Geologic maps, are then used to amplify the motion to obtain surface ground shaking. Surface geology maps, however, often have to be translated to reflect soil classifications used in the empirical attenuation relationships. Then, the soil class map is combined with the bedrock ground motion maps to produce surface ground motion maps.

Figures 6 shows the map of bedrock ground motion acceleration (in g's) for Palo Alto, California from a magnitude 7.5 earthquake on the San Andreas fault (from Kiremidjian et al, 1997). The surface geology map (from Wentworth, 1993) for the same region is shown in Figure 7. The surface peak horizontal acceleration for Palo Alto, California for a magnitude 7.5 on the San Andreas fault is shown on Figure 8.

These maps illustrate how GIS can be used in a simple way to develop macrozonation maps. Ideally, it would be desirable to combine such macrozonation with more local information. Thus, as high hazard areas are identified based on the larger scale maps, a point and click operation can be imbedded in a GIS tool to enable to used to perform more detailed analysis at the specified location. Data, such as, borehole or slope or depth to ground water can be retrieved and viewed before analysis is performed, enabling the analyst to judge the need for further analysis. These are only a few examples of how various levels of analysis can be performed within a GIS environment.

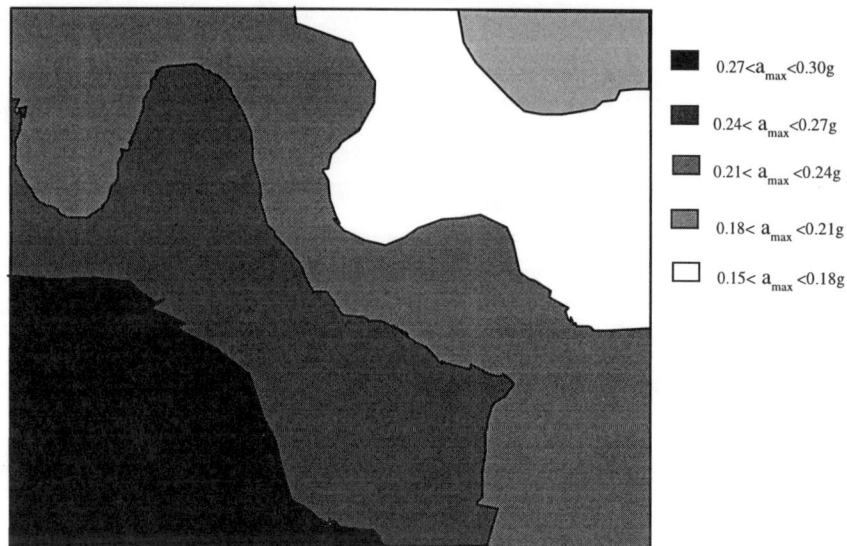

Figure 4. Map of maximum ground surface acceleration from DESRAMOD-2
(Isvandiar, Doroudian and Vucetic, in Kiremidjian et al., 1997).

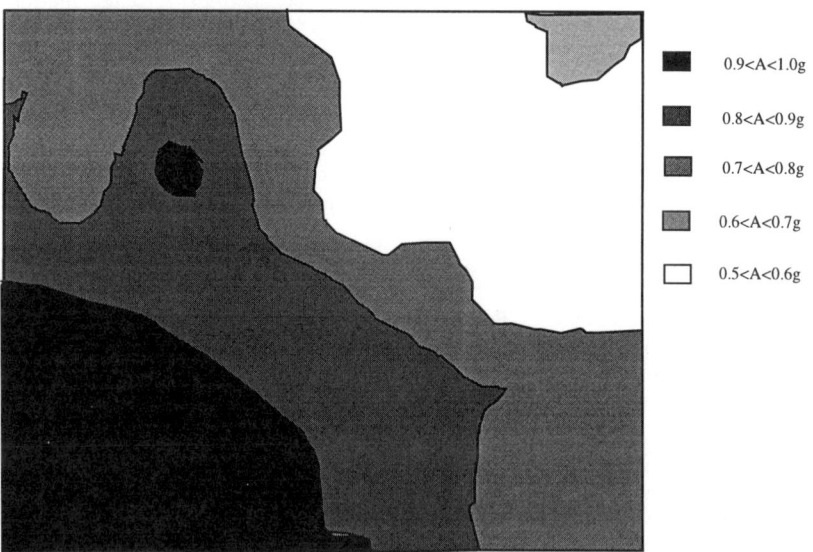

Figure 5. Map of ground motion amplification/deamplification obtained from
nonlinear site response analysis
(Isvandiar, Doroudian and Vucetic, in Kiremidjian et al., 1997).

6

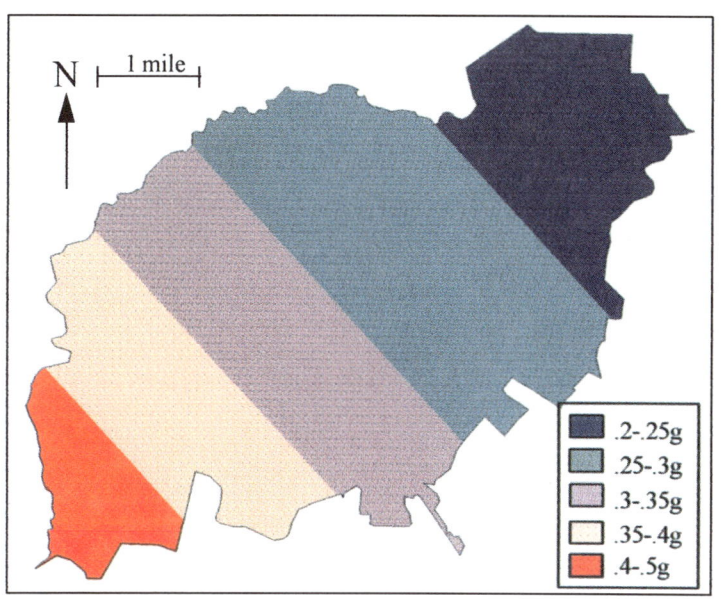

Figure 6. Peak horizontal bedrock shaking in Palo Alto, California for a magnitude 7.5 earthquake on the San Andreas fault (Kiremidjian et al., 1997).

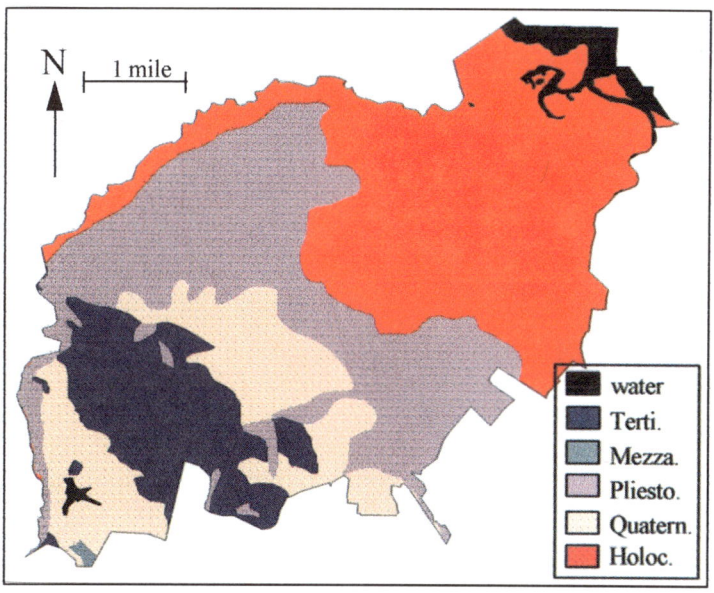

Figure 7. Surface geology map for Palo Alto, California (Wentworth, 1993).

Liquefaction Hazard Mapping

Similar to ground motion amplification studies, liquefaction hazard mapping can also be done on the local or a regional basis. In this paper, only the regional liquefaction potential is addressed. For macrozonation purposes, two parameters are mapped - the *liquefaction susceptibility* and the *amount of ground deformation*. The information required for liquefaction analysis includes *surface geology, depth to ground water, surface ground shaking* and *magnitude* of earthquake.

Liquefaction susceptibility is most frequently characterized according to the classification developed by Youd and Perkins (1978). In order to display the liquefaction susceptibility on a map, the probability that liquefaction would occur is defined as the proportion of map unit that might liquefy. Power et al. (1992) developed five liquefaction susceptibility levels based on the proportion of map unit that is likely to liquefy. Their approach was used in the National Institute for Building Sciences study on earthquake hazard evaluation (RMS, 1995). Using their relationships, the likelihood of liquefaction is assumed to be affected by the ground shaking amplitude, earthquake magnitude and depth to ground water according to the following equation

$$P[liquefation_{sc}] = (P[liquefaction_{sc} | PGA = a] P_{ml}) / (K_M K_W) \qquad (1)$$

where:

$P[liquefaction_{sc} | PGA = a]$ = conditional probability of liquefaction for a given ground surface acceleration for a specified susceptibility class;
P_{ml} = proportion of map unit susceptible to liquefaction
K_M = moment magnitude correction factor defined by equation 2.
K_W = depth to ground water correction factor defined by equation3.

The correction factors for magnitude and depth to ground water are given by Seed and Idriss (1982) as follows:

$$K_M = 0.0027 M_W^3 - 0.0267 M_W^2 - 0.2055 M_W + 2.9188 \qquad (2)$$

$$K_W = 0.0227 d_W + 0.93 \qquad (3)$$

where M_W is the moment magnitude and d_W is the depth to ground water in feet.

This approach was used in the development of liquefaction susceptibility map for Palo Alto, California. The average depth to ground water was generated from the borehole data gathered form the study region. The scenario event was assumed to be

8

a 7.5 on the moment magnitude scale. Using the surface geology map, the geologic units in the region were assigned to a liquefaction susceptibility class. The surface ground shaking map for magnitude 7.5 earthquake on the San Andreas fault shown in Figure 8 provide the peak horizontal acceleration values required for the analysis. Figure 9 shows the liquefaction susceptibility map for Palo Alto (Kiremidjian et al., 1997) generated using equation. Based on this map, more advanced liquefaction studies can be performed for the high liquefaction potential areas.

For engineering design and site remediation purposes, it is not sufficient to predict only the liquefaction potential. It is also necessary to estimate the permanent ground deformation that can result from liquefaction. The expected permanent ground deformation is given by:

$$E[PGD_{sc} = K_\Delta E[PGD|(PGA / PL_{sc}) = a] \qquad (4)$$

where

$E[PGD|(PGA / PL_{sc}) = a]$ = the expected permanent ground displacement for a given susceptibility class under a specified level of normalized ground shaking;

(PGA / PL_{sc}) = empirical relationship developed by Youd and Perkins (1978) and modified by Sadigh et al. (1986);

K_Δ = displacement correlation factor developed by See and Idriss (1982) and is given by equation 5.

$$K_\Delta = 0.0086 M_W^3 - 0.091 M_W^2 - 0.469 M_W + 0.9835 \qquad (5)$$

where M_W is the moment magnitude.

These models are based several empirical relationships using data from past earthquakes. Their application on regional level is relatively simple because they do not require extensive amount of local soil parameter data which typically are not available nor do they require a major computational effort that needs to be repeated at many points on a map. In contrast, advanced analytical approaches that include the nonlinear behavior of soils and the dynamic characteristics of the ground motion and soil cannot be used at present for macro or microzonation. Most frequently these models require that local soil parameter data be available and that ground motion time histories be generated at each site. Although much has been learned from recent earthquake event about the liquefaction mechanism and considerable additional data has been obtained, there is a clear need to improve the simple empirical approaches for liquefaction potential and ground deformation mapping.

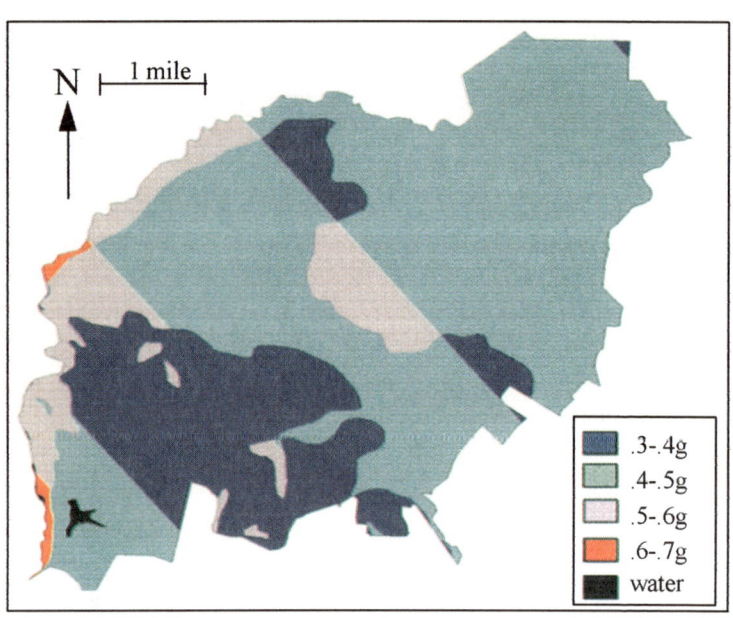

Figure 8. Peak horizontal acceleration in g's at the ground surface level for Palo Alto, California for a magnitude 7.5 earthquake on the San Andreas fault (Kiremidjian et al., 1997).

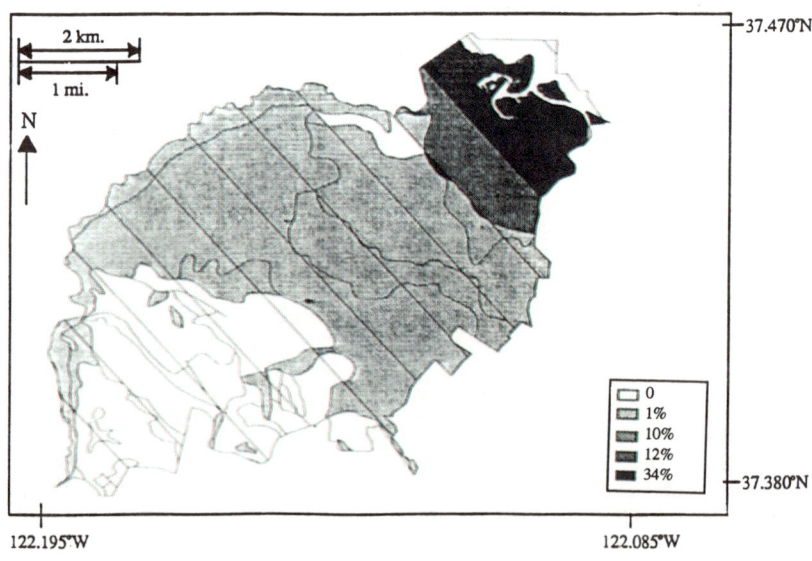

Figure 9. Probability of liquefaction in Palo Alto, California for a magnitude 7.5 earthquake on the San Andreas fault (Kiremidjian et al., 1997).

Landslide Hazard Modeling

Earthquake induced landslide hazard analysis requires that the potential for landslide occurrence, amount of ground deformation and area of sliding be defined. The most widely used methods for earthquake induced landslide hazard analysis are based on the simple pseudostatic approach (Wilson and Keefer, 1985) or on the modified Nemark method (Jibson, 1992). Wilson and Keefer (1985) provide landslide susceptibility definitions as functions of the slope angle, geologic unit, and the seasonal water content (dry or wet). For each geologic unit, Wilson and Keefer also define lower bounds for the slope angle and critical acceleration below which landslide will not occur. Furthermore, they define values of critical acceleration required to induce landslide for different landslide susceptibility classes.

Landslide mapping methodologies include generation of slope, local geology, and geotechnical parameter maps. Figure 10 shows the main components of such a methodology.

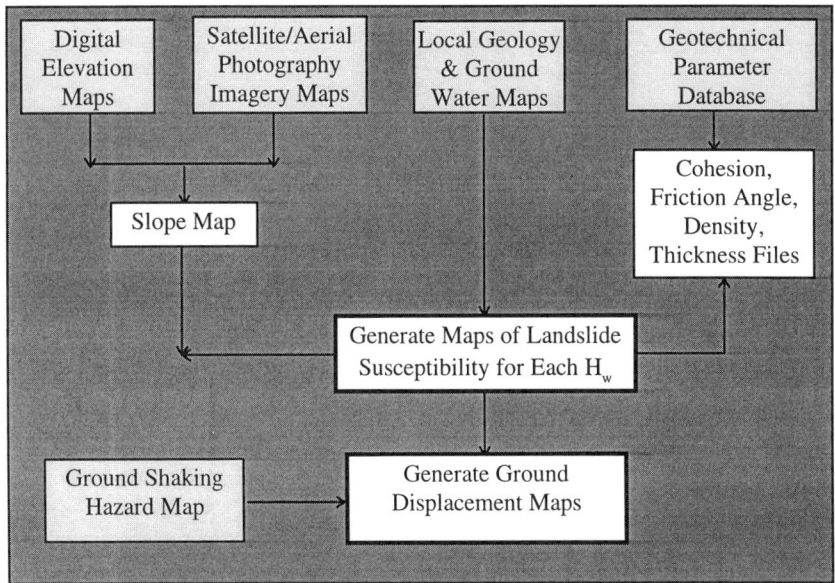

Figure 10. Methodology for landslide susceptibility and ground displacement mapping.

As an example the elevation map from a digital line graph (DLG) at 1:100,000 scale for a small area in Palo Alto, California is shown on Figure 11. The slope map

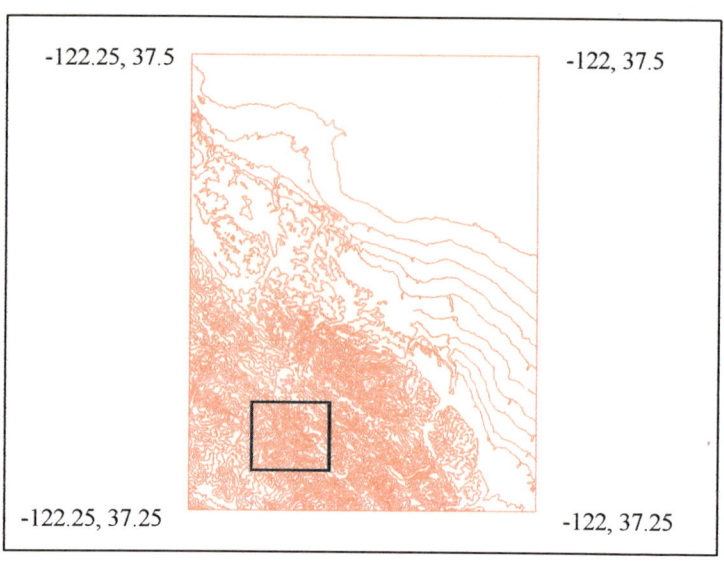

Figure 11. Ground surface elevation map for section of Palo Alto, California obtained from a 1:100,000 Digital Line Graph (from Hofmeister, 1997).

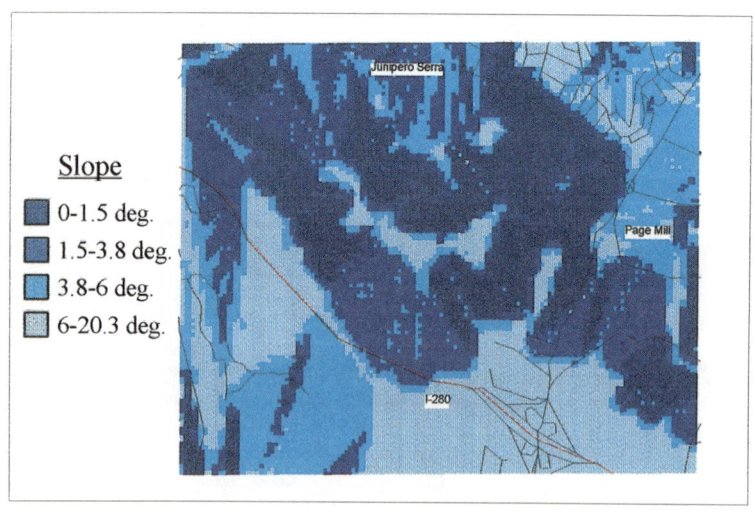

Figure 12. Slope angle map for an area within Palo Alto, California (from Hofmeister, 1997).

generated at a 25 m grid scale is shown on Figure 12 (Hofmeister, 1997). These maps can then be utilized to generate landslide susceptibility and ground deformation maps. The generation of slope angle maps can be challenging. The slope map shown on Figure 12 was generated in a 3D GIS ($3D$-$Maps^{TM}$) to facilitate this process. Depth to ground water maps are, in general, difficult to obtain and can very based on seasonal changes or drought/rainy period cycles that may span several years.

At present there are no methods for estimating the area over which a landslide will take place. Conceptually, the area of sliding for a given critical slide angle will be the intersection between the plane along that critical slide angle and a slope map. Practically, however, this is not a simple problem to resolve.

Conclusions

Mapping of ground motion amplification, liquefaction and landslide hazard were presented in this paper within the framework of geographic information systems. Examples application for each of these hazards were shown primarily for the Palo Alto, California. The paper focused on regional hazard analyses, although, an example for ground motion amplification on a more local scale for which data were available was also included. From these examples, it can be observed that some of the greatest difficulties lie in the generation of the basic data needed for the different types of analysis. These data, however, are increasingly becoming available making it GIS tools more and more useful. An important aspect in these analysis is the assessment of uncertainties in the various data bases and analytical models. Currently, these uncertainties are not treated in any manner and should be the focus of future studies.

Acknowledgments

I would like to thank Dr. Stephanie King and Mr. Jonathan Hofmeister for generating the maps for Palo Alto, California. We are indebted to Environmental Systems Research Institute for making the software ARC/IFO^{TM} and to Minesoft Corporation for making the $Techbase^{TM}$ software available for our use. Much of the support for this work came from the California Universities for Earthquake Engineering Research (CUREe) under the CUREe/Kajima Phase II projects.

References

Boore, D. M., Joyner, W. B., and Funal, T. (1993). "Estimation of Response Spectra and Peak Ground Acceleration from Western North American Earthquakes: An Interim Report." Open File Report 93-509. United States Geological Survey, Menlo Park, CA.

Dorooudian, M., Vucetic, M., and Martin, G. (1995). "Development of Geotechnical Database for Los Angeles and Its Potential for Seismic Microzonation."

Proceedings of the 5th International Conference on Seismic Zonation, Nice, France, pp. 1514-1520.

Hofmeister, J. (1997). "Development of Slope Maps in 3D-GIS." K2 Technologies, Inc. Internal Report, San Jose, CA.

Iskandar, V., Doroudian, M. and Vucetic, M. (1996). "Development of Geotechnical Database for Palo Alto and Its Utilization for Microzoning." *Proceedings of the 11th World Conference on Earthquake Engineering*, paper No. 1277, Acapulco, Mexico, July.

Jibson, R. W. (1992). "Predicting Earthquak-Induced Landslide Displacements Using Newmark's Sliding Block Analysis," Transportation Research Record 1411, U.S. Geological Survey, Denver Federal Center, Denver, CO.

Kiremidjian, A. S., King, S. A., Law, K., Basoz, N., Singhal, A. Straser, E., Belubekian, M., Moehle, J., Olson, R., Goettel, K., Eidinger, J., Horner, G., Vucetic, M., Doroudian, M., and Iskandar, V. (1997). "Methodologies for Evaluating the Socio-Economic Consequences of Large Earthquakes." Final Report to CUREe, for CUREe-Kajima Phase II Projects.

Matasovic, N. and Vucetic, M. (1993). "Seismic Response of Composite Horizontally-Layered Soil Deposits." Research Report, Civiland Environmental Engineering Department, University of California at Los Angeles, Los Angeles, CA.

Power, M. S., Dawson, A. W., Streiff, D. W., Perman, R. G. and Haley, S. C. (1982). "Evaluation of Liquefaction Susceptibility in San Diego, California Urban Area." Proceedings of the 3rd International Conference on Microzonation, Vol. II, pp. 947-968.

Sadigh, K., Egan, J. A., and Youngs, R. R. (1986). "Specification of Ground Motion for Long Period Structures." *Earthquake Notes*, Vol. 57., No. 1, p.13.

Seed, H. B., and Idriss, I. M. (1982). "Ground Motion and Soil Liquefaction During Earthquakes." Earthquake Engineering Research Institute, University of California at Berkeley, Berkeley, CA, 134p.

Wilson, R. C., and Keefer, D. K. (1985). "Predicting Areal Limits of Earthquake Induced Landsliding." *In Evaluating the Earthquake Hazard in the Los Angeles Region- an Earth Science Perspective*, Professional Paper 1360, U.S. Geological Survey, pp. 316-345.

Youd, T. L. and Perkins, D. M. (1987). "Mapping Liquefaction Severity Index." *Journal of the Geotechnical Engineering Division*, ASCE, Vol. 118, No. 11, pp. 1374-1392.

Spatial Geologic Hazard Analysis in Practice

J. David Rogers[1], Member, ASCE

Abstract

This article seeks to focus on applications of Geographic Information Systems (GIS) to geologic hazard assessment, as developed in actual consultations in the San Francisco East Bay over the past decade. The applications examined include acquisition of published geologic data onto GIS, fabrication of new products using GIS technologies, combining political concerns and legal restrictions with geohazards into new composite GIS products, and preparation of detailed geohazard maps for earth movement potential. The last areas evaluated are three dimensional storage and retrieval of subsurface geologic information, with emphasis on providing a means to assess ground water resources and expected seismic site response. A number of problems inherent to GIS representations of voluminous geologic data are then discussed in the conclusions, forewarning readers of the many limitations not commonly appreciated by end users of GIS products.

Introduction

Most GIS products build upon the familiar format of spatial maps, commonly presented in the form of recognizable cadastral base maps, such as highway, parcel or topographic maps. Today most land survey products are produced on orthophoto bases, in a digitized format easily applied to any range of GIS software. An orthophoto base map greatly enhances accuracy in field-locating, and the recent emergence of inexpensive Global Positioning System (GPS) receivers has greatly aided accurate field locating.

For those practicing in the applied earth sciences, the GIS product portends some recognizable liabilities, in that a two-dimensional (planar map) representation may fail to convey an accurate picture of geologic risk, particularly since such risks often lie

---

[1] Principal, Rogers/Pacific, Inc., 396 Civic Drive, Pleasant Hill, CA 94523-1921

beneath the ground surface, while maps only portray what outcrops. GIS also affords the fabrication of politically-fashioned map products, a new application whose ramifications have not yet been fully appreciated, and within which end users could be unduly influenced to draw erroneous or incomplete conclusions. Today, most GIS products are presented in an attractive format, with beautifully orchestrated computer graphics. The use of aesthetic graphics emanating from authoritative sources, such as government agencies and research institutions, are willingly accepted by end users as a "last word", seemingly without fault or blemish.

The balance of this article seeks to explore some of methods employed and focuses on unintentioned problems associated with GIS products in educating end users, such as engineers, scientists, educators, politicians and the public. GIS is here to stay, but as a profession we need to develop a cognizance of it's limitations and liabilities. Four GIS geohazard products released in the San Francisco Bay area are briefly profiled, exploring their strengths and weaknesses.

Moraga Development Capability Map (1988-89)

This study was undertaken as a joint project with the Spatial Information Systems Laboratory at U.C. Berkeley's Center for Environmental Design Research in 1988 (Rogers/Pacific, 1989). The GIS database used was the GRASS (Geographic Resources Analysis Support System) software developed by the Army Corps of Engineers' Construction Engineering Research Laboratory. The project commenced by tabulating 21 physical attributes within the Townof Moraga, an area of approximately 24.6 km², taken mostly from existing sources of data. Some of the basic data attributes included: elevation, slope, slope aspect, land use categories, parcels, bedrock geology, faults and folds, landslides, FEMA flood hazard, streams and swales, soils, soil shrink-swell potential (from SCS), Storie Index (from SCS), ridge lines, distribution of colluvium and vegetation.

The GRASS GIS program was used to superpose a grid over the polygon map data and reconfigure the data so that each map overlay consisted of grid squares. A grid size of 15 meters square was used, the smallest resolution felt appropriate to a General Plan level of analysis (a finer grid could have been used). The strength of the grid method of storing data maps is its capability for analysis using varying combinations of overlays. Map layers could then be multiplied, divided, added to one another, or otherwise manipulated in order to create the desired map product. The simplest operation was reclassification, in which two existing map categories are assigned some specified attribute. For example, a soil erosion potential map was created by specifying Soil Conservation Service (SCS) erosion hazard ratings for each soil type on the standard County-wide SCS maps. In addition, distance buffers could be provided around recognized hazards, such as along scenic ridge lines, the toes of swales prone to debris flows, or along the crest of steep creek banks, in order to conform with existing planning and safety restrictions.

The Town contracted for the GIS system in order to facilitate the creation of a planning document, called a "Development Capabilities Map". The purpose of this map was to provide a spatial update to their General Plan, necessitated by the recent creation of a controversial open space ordinance approved by the voters in 1986. Five other maps showing open space ordinance restrictions and two displaying development prohibitions were also prepared for review by the Town Council. The council selected six physical attributes felt most important: Ridge lines, Landslide Susceptibility, Slope, Flood Hazard, Vegetation and Soil Erosion. For each of the selected factors a 10-point scale was developed, 10 being most restrictive and 1 being least. For example, being on an active landslide or atop a restricted ridge line would equate to a "10,", while simply being within 100 feet of an active landslide would only be a "6." In this way, both geologic hazards, graded by scientists, and legal/political concerns, such as ridge line development, could be weighted according to concern and combined to form a composite map product.

The end product was a "Development Capabilities Map" (Fig. 1), essentially a politically-fashioned document, based upon physical geologic input (including topography). The map utilized warm and cool colors, arranged upon the 15m² grids. Warm colors (red-yellow-green) denoted areas most suitable for development, while cooler colors (blue-purple-magenta) denoted areas least suitable for development. In this way, citizens were afforded the opportunity to voice their concerns about various geohazards, politicians could incorporate public opinion, and the Town's legal counsel outlined the legal restrictions imposed by the open space ordinance. These factors were combined with the geologic input to fabricate a planning document in a map form. In order for the map to be legally binding it had to apply equally to all portions of the Town and survive rigorous public review and commentary. The plan has now been in force a little over eight years and development applications continue to be promulgated through the Town using the Development Capabilities Map as a controlling document.

1993 Orinda Landslide Mapping Program

This project sought to map landslides and erosion hazards in Orinda, California, an area of 33.15 km². Like nearby Moraga, Orinda lies within the steeply carved East Bay hills of the San Francisco East Bay area. Mostly developed between 1946-66, it is an area that has long been recognized for spawning numerous landslides, many within a few years of hillside grading for development (Kachadoorian, 1956; Kirsch, 1960; Radbruch and Weiler, 1963; Nilsen and Turner, 1975; Duncan, 1971). A detailed report was prepared for the city to accompany the landslides and surficial deposits hazard maps (Rogers/Pacific, 1994).

In the late 1970s the USGS had pioneered the use of computerized mapping of landslides hazards in San Mateo County (Newman, Paradis and Brabb, 1978). However, this study was the first opportunity to prepare a slope stability hazard product at a large enough scale (1:3,600) to be property-specific; in other words, large

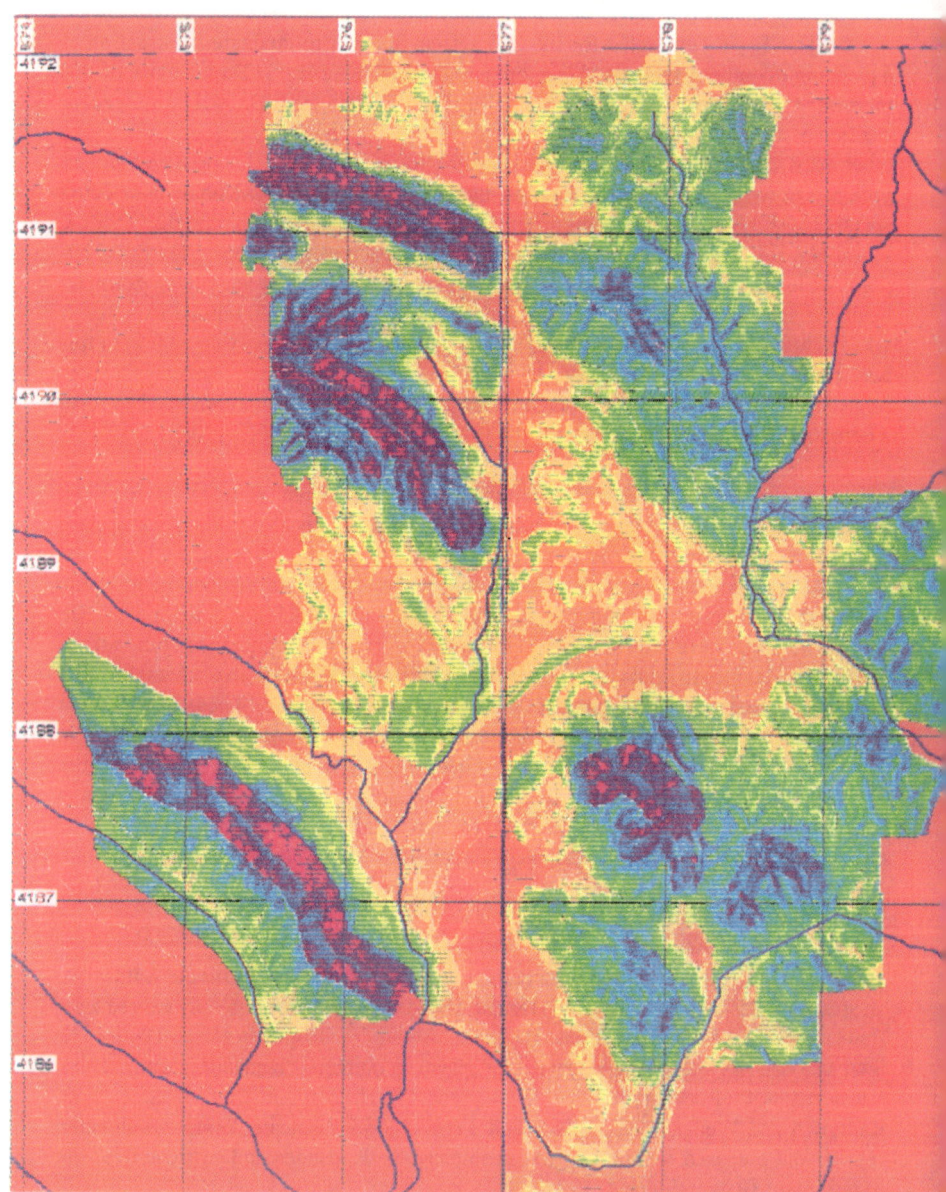

Fig. 1 - GIS generated Development Capabilities Map prepared for Town of Moraga, California in 1989. The map was based upon the addition of six factors, all weighed equally: ridgelines, landslide susceptibility, flood hazard, slope, vegetation and soil erosion.

enough to allow individual parcel owners to easily assess the areal limits of mapped hazards with respect to their dwellings and property lines. An essential part of the study was the clients' desire to create a GIS product which would incorporate the County's tax assessor parcel maps as the base layer, insofar as citizens and government personnel desired an off-the-shelf product that would delineate upon whose properties various hazards were supposed to exist.

From past experience we had found cadastral maps unsuited for geohazards mapping because of the lack of any meaningful ground references, such as dwellings, paved areas, vegetation patterns and topography easily discerned on aerial photo imagery. The decision was made to map landslides and colluvium on orthophoto topographic sheets, then input these line contacts onto a GIS layer using AutoCad 12, because topographic data is normally delivered on AutoCad. The assessor's parcel map on ArcInfo was brought aboard via a DXF file transfer and subsequently replaced the orthophoto topographic map as the reference base map most commonly desired by end users (although other base map products, such as orthophotos and topography can be made available, upon request).

The contour interval chosen for the topographic base map was 1.52 m (5 feet). At this scale slides as small as 8 to 10 meters of total vertical height could be reliably identified on the basis of anomalous topographic keys. Our experience has shown that contour interval is the most important factor in delineating landslide features, even more reliably than stereopair aerial photos. By focusing on those slope areas with anomalous topographic expression, "target areas" for further examination were identified.

The size of discernable landslides features is intimately tied to resolution and contour interval of the base orthophoto topographic map. Only the largest slides can be discerned on standard USGS 1:24,000 (7.5-minute) topographic quadrangles. This is because one needs at least 5 contour intervals across the slide to recognize topographic indicators of landslippage. The most common topographic intervals on USGS products are 6.10 m (20 feet) and 12.19 m (40 feet), meaning slides would need to be 30+ to 60+ meters in vertical relief to be recognized on the basis of just topographic anomalies.

After areas of anomalous topography are identified, the next step is to make careful evaluation of these same areas with sets of stereopair aerial photos. For this study eight sets of stereopair aerial photos covering the study area were examined, dating from 1928 to 1990. Despite the fact that aerial photos have long been recognized as the preeminent method to make reconnaissance-level evaluation of landslide hazards (Liang and Belcher, 1958; Ritchie, 1958), there are also a great number of problems associated with delineating landslides on aerial photos, including tree canopy and slope aspect, parallax distortion, scale resolution, sun angle and soil moisture.

Because it was a reconnaissance-level product covering over 33 km², direct filed observations were not made as part of the study (normally, the third step in the identification process). Intended limitations of the map's accuracy were indicated in a disclosure statement on each sheet, directing users to retain their own consultants to perform site-specific studies if desiring to seek confirmation or denial of mapped units. The last step normally taken in evaluating landslides is site-specific subsurface investigation. Subsurface exploration techniques involved with dormant, ancient or inactive bedrock landslides are often tedious. Common methods of field evaluation include deep trenches across suspected headscarp grabens and large diameter bucket auger excavations in various parts of the suspected mass which can be downhole logged by a geologist.

Unlike the Moraga study discussed previously, this effort sought to summarize the surficial hazards into seven basic map categories. The types of surficial deposits included: 1) alluvium; 2) stream terrace deposits; and 3) colluvium, or slope wash, deposited within old bedrock ravines. Landslides were divided into four major categories, those being: 1) debris flows (including source and runout areas); 2) Earthflows; 3) Translational-Rotational bedrock landslides; and 4) Ancient or Indistinct Landslide deposits. An example of the end product is shown in Fig. 2. Boundaries of mapped units were shown in conventional nomenclature, with solid lines denoting the aerial limits of recently active slide deposits; dashed lines delineating inferred limits; and dashed with query where contacts were concealed.

1991 Study of the Geology Underlying the Oakland-Alameda Area

Following the October 1989 Loma Prieta earthquake, geoscientists were desperate to access geologic data regarding the late Quaternary-age stratigraphy and geologic structure beneath the central San Francisco East Bay plain. Most existing data was unpublished and scattered in a wide range of formats, difficult to track down and correlate. Of greatest interest was information from deep wells, sometimes penetrating 400m or more of late quaternary age sediments into the underlying Jurassic-Cretaceous age basement rocks of the Franciscan assemblage (Rogers and Figuers, 1991).

We managed to collect the logs of over 200 borings between 37 and 316 meters deep within the study area, ranging in age back over 100 years. For older wells, location references can be extremely difficult to verify, as most of the referenced land features no longer exist. Reliable maps from the era in question were then accessed in order to verify the described positions. It soon became apparent that water well drillers had been notoriously careless in stating precise locations of their wells, and some well heads were found upwards of a kilometer from their described positions.

One of the problems inherent in compiling subsurface geologic data from a wide range of sources is to accurately account for variances in interpretation and historical

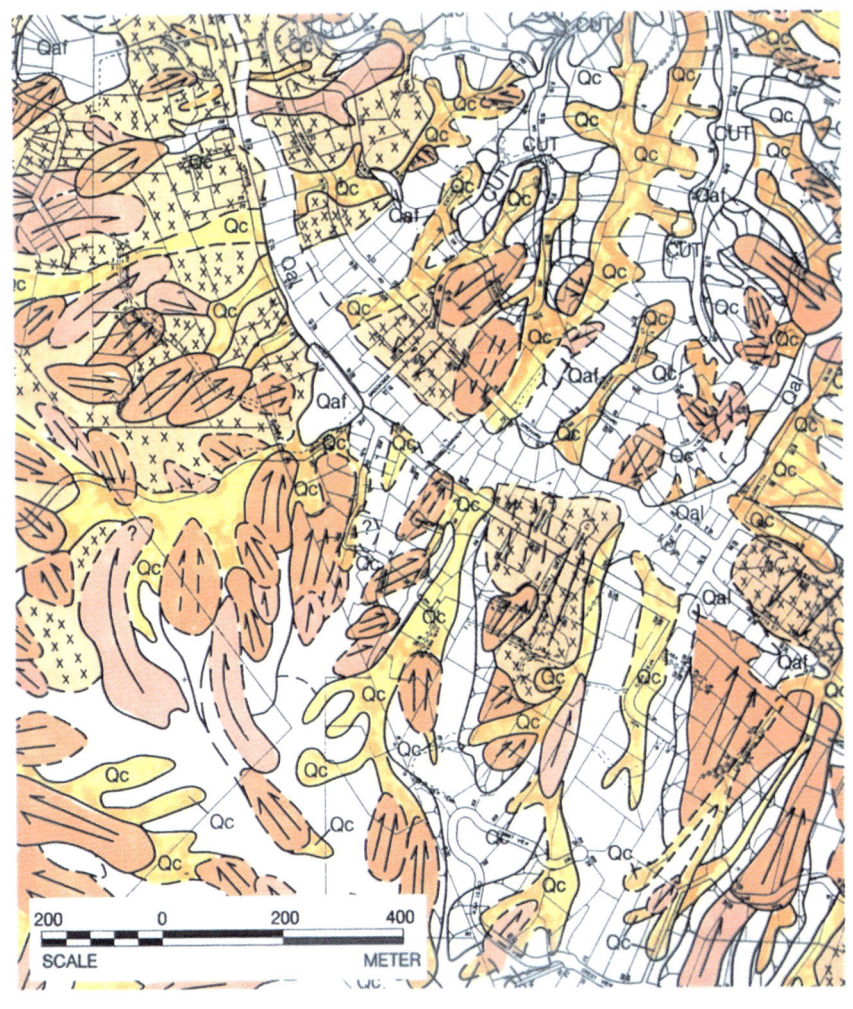

Fig. 2 - Interpretive landslide features overlain on assessor's parcel map, showing a portion of Orinda, California, originally produced at a scale of 1:3,600. Arrows indicate direction of supposed movement, while check-marked areas are those supposed to be underlain by old bedrock landslides.

evolution of the stratigraphic nomenclature. Two steps are needed before meaningful correlations can begin: an accurate idea of the bedrock basement geometry, which controls overall sediment thickness; and a coherent representation of the sediment stratigraphy, or the vertical sequence of units previously identified by all sources.

The San Francisco Bay bedrock depression turned out to be structurally controlled, by the existence of right-lateral strike-slip faults (the San Andreas and Hayward) bounding the Bay, but a pull-apart basin beneath the central portion of the East Bay plain was hypothesized to explain the gravity anomalies and much deeper accumulation of late quaternary age sediments occupying this area southeast of Oakland (Fig. 3). Further spatial analysis revealed that less than 5% extension was necessary to explain the pull-apart basin, and that a maximum depth to Franciscan basement of something less than 500 m could be expected (Fig.3) .

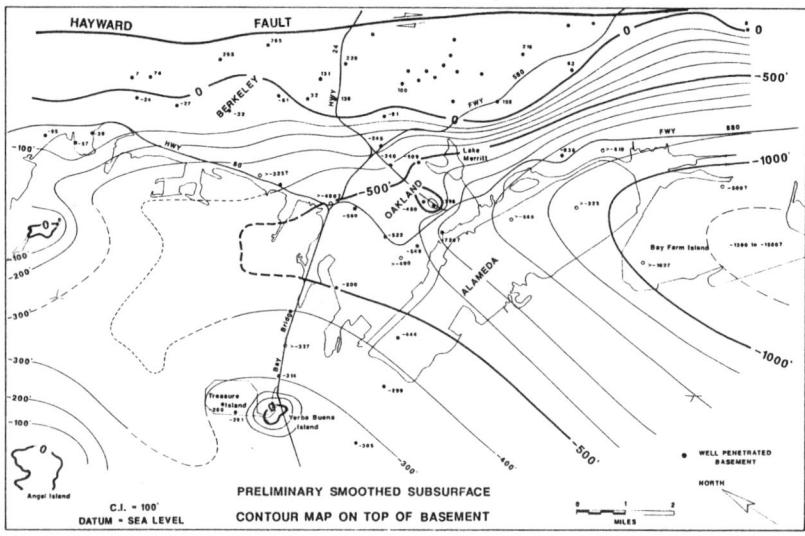

Fig. 3 - Preliminary smoothed subsurface contour map on top of Jurassic-Cretaceous age Franciscan assemblage basement in east central San Francisco Bay. Those wells which penetrated the basement are shown as solid dots. Estimates of expectable site response could be generated from such data, provided shear wave velocity profiles of the overlying quaternary-age fill were available.

Once the overall structural geologic setting was firmly established, a conceptual model of the stratigraphy was formulated. Geotechnical well logs generally utilize descriptions of physical attributes, without any serious consideration of stratigraphic unit assignment and faunal age, which is usually derived from detailed evaluation of micro fossils (Sloan, 1992). We were left with logs that repeatedly described "bay mud," "blue marine clay," or "estuarine mud," without any reference to relative age and

22

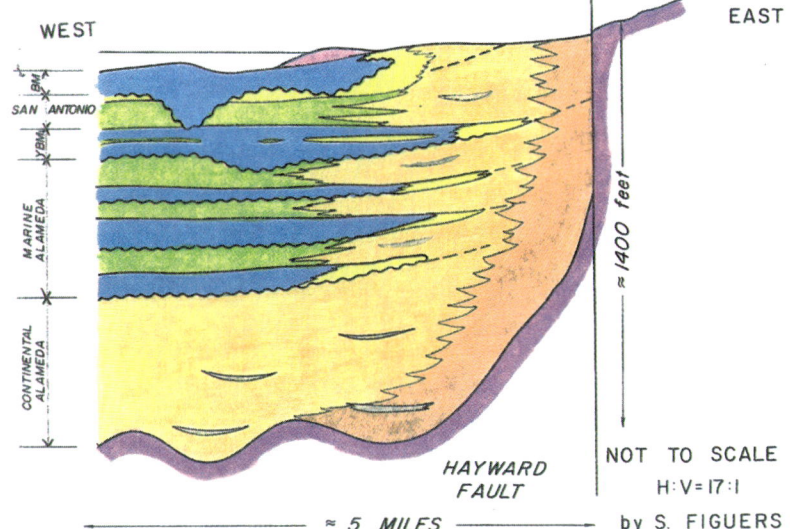

Figure 4 - Schematic section view (with 17:1 vertical exaggeration) of typical stratigraphic relationships across the eastern shoreline of central San Francisco Bay. At least three, and possibly four, landward transgressions of late Pleistocene seas are recorded in the upper Alameda formation. Appreciating the sequence of depositional events was key to making correlations between well logs of dissimilar age and origin.

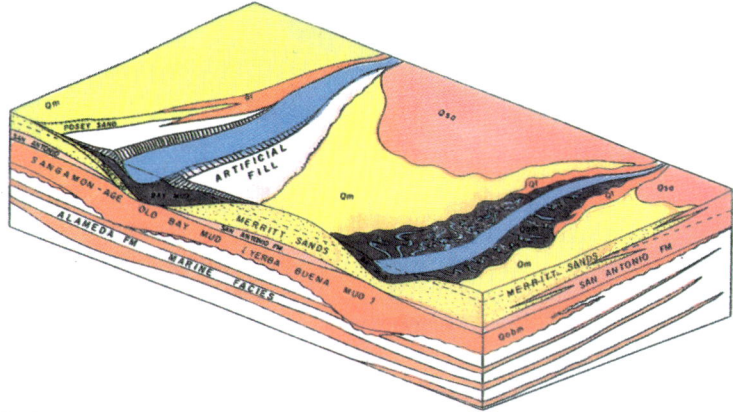

Fig. 5 - Schematic block diagram view looking north across the present East Bay shoreline in vicinity of Oakland and Alameda. Along the Bay margins, bay mud has been deposited in Wisconsin-age channels, cut into the San Antonio formation when sea level was 107m lower than today. It is important to appreciate the lateral restriction of channel deposits when attempting to correlate well logs. Note how succeeding channels are not always situated one atop another, but may be offset.

23

position. By carefully compiling the best information from the deepest and most complete well logs, we eventually developed what we believed to be an accurate model of the underlying stratigraphy, shown in Fig. 4.

A reliable model of the stratigraphy is critical to understanding what sequences of sediments we could expect to see at any given locale. We were then able to use "sequence models" in lieu of micro fossils from actual cores to unravel likely unit assignments and relative ages. A sequence model is something akin to a fingerprint at any given locale. Major weather pattern and base level changes are faithfully recorded in the Bay's sediments (Sloan, 1992), and it is the sequence, or vertical order of the various sediments, that attests to sea level encroachment or regression during interglacial periods. By evaluating the entire late quaternary stratigraphic column, an experienced stratigrapher could usually discern accurate age and unit assignments for the various sediment sequences described in the old well logs.

One of the most useful products from the study was a better understanding of the three-dimensional aspect of superposed units, a factor often ignored by many geotechnical professionals, because they usually work on relatively small parcels. As shown in Fig. 5, channel deposits were found to be laterally restricted, seldom extending far in a direction transverse to the old channel axis. Most of these channels support paleo flood plains, within which lie overbank silts. Though commonly referred to as estuarine muds, in 75% of the cases analyzed, these materials turned out to be overbank silt, deposited at flood stage. The block diagram in Fig. 5 shows how individual units interfinger and can be locally missing, causing many an incorrect interpretation. The complexities of working in low gradient, fresh water-salt water mixing environments, cannot be understated.

1993-95 USGS Well Repository Project

This project, funded by the U.S. Geological Survey , sought to follow up on the well database begun in the 1991 NSF study, but expanded to the entirety of Alameda and Contra Costa Counties, a land area of 3,806 km^2 . The goal of the project was to establish a computerized database methodology by which meaningful data could easily be retrieved from voluminous quantities of existing well data, mostly in the form of water wells. Some 4,700 well logs were evaluated and tabulated into the database, using the program Paradox 4, which had previously been exploited by petroleum firms to categorize their well data. An example of a graphic print out is presented in Fig. 6.

In California the State Department of Water Resources has collected the logs of all water wells for over 100 years. Though not accessible to the public at large, these logs are available to researchers working with government agencies. Approximately 3,200 well logs were provided from this depository for inclusion in this study. A common problem with using water wells is their poor well log descriptions, which can be almost uselessly vague. As a consequence, a major goal of this study

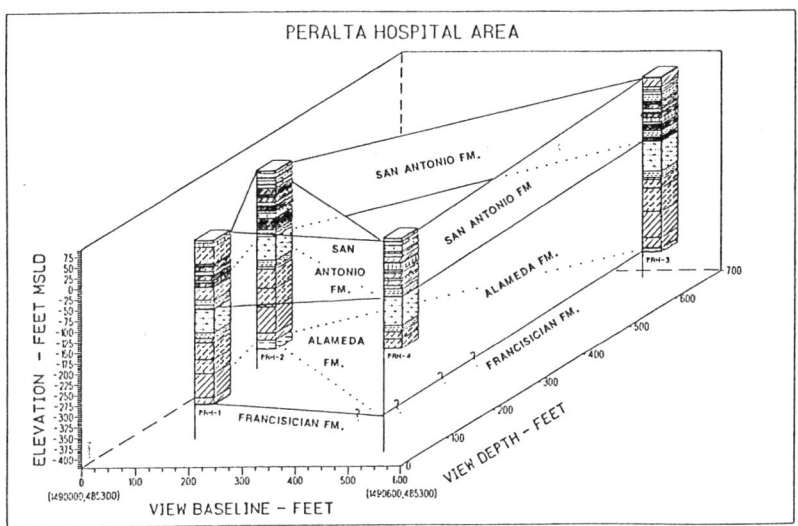

Fig. 6 - An example of GIS geologic data base manipulations made with the program GTGS, using data taken from the USGS well repository, stored on Paradox 4. The example presented is a stratigraphic fence diagram that allows detailed cross-hole correlations of subsurface data.

was to create a means by which quality data could be sifted from large volumes of low quality data. Though the goal of most any GIS study, this is easier said than done.

On a GIS map users will ASSUME that all the input data is of like kind and quality. In this project we were dealing with well data that extended over 100 years from hundreds of independent sources. During the time interval covered, there were innumerable variances in stratigraphic nomenclature, not to mention natural variations in interpretation. Just how great these variables were was appreciated when comparing the logs of adjacent wells drilled and logged by different agencies and methods dozens of years apart.

As the GIS product provider, we were then faced with the dilemma of how to store all the available data, but provide a means by which end users could evaluate the comparative quality of the various logs. Most end users favor data generated within their respective professions, i.e. geotechnical engineers favored geotechnical boring logs, while petroleum industry users favor logs generated within their industry. We eventually decided that we could not "grade" the various entries in the database, we had to trust users to make reliable interpretations, seeing raw well logs for what they really are, limited information, oftentimes little better than nothing.

The final product (Rogers, Figuers and Mero, 1994) contained 43 registers of information, including a battery of location identifiers, such as latitude, longitude,

25

California Well Coordinate System, etc. Other registers included information on well depth, and all manner of log annotations, such as: lithologies encountered, geophysical data, water table information, casings, penetration resistance, sampling, testing, and even the type of well seals applied. Because of the great number of information registers, end users could query the database for data of particular interest, such as: which wells penetrate the Franciscan basement and have electric logs? The data base would then scan for those wells which met each of the requested criteria and answer.

A GIS well database is not difficult to set up with raw data, provided such data can be retrieved. However, it is a time consuming and tedious process to estimate well collar locations, which need to be verified or the spatial nature of the database would be erroneous. As a consequence, data input can involve substantive labor. Computerized well databases are the only available method by which useful information can be skimmed from large volumes of data. The GIS representation of such data, spatially arrayed is essential to formulating any meaningful conclusions about underlying geologic structure.

Conclusions

Existing GIS databases can store enormous quantities of geotechnical data, allowing for unprecedented libraries of information available for manipulation that were impossible to contemplate just a decade ago. The spatial representation of data on map-style layers is invaluable as a tool of communication with the end users of such information, but such products are also fraught with limitations we should not lose sight of. These include:

1.   *An inherent problem with this form of spatial representation is the inability to inform users of what might lie immediately beneath the ground surface.* End users evaluating two-dimensional GIS-derived map products without sufficient training in geology may draw erroneous conclusions from such maps, by assuming that the depicted surface deposits extend to some unknown depth. In the interpretation of landslide features, this problem is particularly acute, insofar as only the most recent generation of slippage is evident to most interpreters. Previous sequences of movement tend to become increasingly masked with time, as one manner of landslippage rests upon another, making detection of superposed events extremely difficult.

2.   *Another problem is the variable quality and sources of data represented on various layers of GIS products.* Whenever we present a set of data with equal line weights or colors, end users will intuitively assume all the data is of like reliability, when in fact it seldom is. An example of this problem was the uncertainty associated with collar locations of old wells, discussed earlier. Some parts of the maps are going to have better data than others. How do we let our

users know the degrees of reliability? This facet of GIS products needs further development.

3. *Like any geologic product, GIS databases will require periodic updating, as new information and interpretative models become available.* The publication of GIS-derived documents by credible sources tends to be seen by end users with a stamp of approval unchanging with time. More often than not, earth science professionals in the private sector will continue to use outdated maps, long after they have been superseded by newer products. By posting future GIS products on the Internet, much of this laziness and familiarity with hard copy products will be avoided, as users will tend to pull down information off the net when they need it, and the requirement for hard copy storage will diminish. However, will we see the funds made available to periodically maintain the GIS database products we produce? We must assume that despite our best efforts, our products will likely become dated, and we should consider the inclusion of disclaimers on all of our work to inform naive end users of the product's limitations, especially in regards to timeliness.

References

Duncan, J.M., 1971, Prevention and Correction of Landslides: 6th Annual Nevada Street and Highway Conference, Section II, April 7, pp. 1-42.

Kachadoorian, R., 1956, Engineering Geology of the Warford Mesa Subdivision, Orinda, California: U.S. Geological Survey Open File Report, 13 p., 1 plate.

Kiersch, G.A., 1969, Pfeiffer vs General Insurance Corporation-Landslide Damage to Insured Dwelling, Orinda, California, and Relevant Cases: in Kiersch, G.A. and Cleaves, A.B., eds., Engineering Geology Case Histories Number 7: Geological Society of America, pp. 81-94.

Liang, T., and Belcher, D.J., 1958, Airphoto Interpretation, in Landslides and Engineering Practice: National Research Council, Highway Research Board Special Report No. 29, pp. 69-92.

Newman, E.B., Paradis, a. R., and Brabb, E.E., 1978, Feasibility and Cost of Using a Computer to Prepare Landslide Susceptibility Maps of the San Francisco Bay Region, California: U.S. Geological Survey Bulletin 1443, 27 p., 1 plate.

Nilsen, T.H., and Turner, B.L., 1975, Influence of Rainfall and Ancient landslide Deposits on Recent Landslides (1950-71) in Urban Areas of Contra Costa County, California: U.S. Geological Survey Bulletin 1388, 18 p, 2 plates, map scale 1:62,500.

Radbruch, D.H, and Weiler, L.M., 1963, Preliminary Report on Landslides in a part of the Orinda Formation, Contra Costa County, CA: U.S. Geological Survey Open File Report, 35 p.

Ritchie, A.M., 1958, Recognition and Indentification of Landslides, in landslides and Engineering Practice: National Research Council, Highway Research Board Special Report 29, pp. 48-68.

Rogers, J.D., Figuers, S.H., and Mero, W.B., 1994, Development of a Computer Well Data Base for Alameda and Contra Costa Counties: final report to U.S. Geological Survey, Contract 1434-92-C-50023, 61 p., 1 plate.

Rogers, J.D., Figuers, S.H., 1991, Engineering Geologic Site Characterization of the Greater Oakland-Alameda Area, Alameda and San Francisco Counties, California: final report to National Science Foundation, Grant No. BCS-9003785, 52 p., 2 plates.

Rogers/Pacific, 1989, Summary Report, Moraga Development Capability Map, Moraga, California: unpublished consultant's report for Town of Moraga, CA, job number PEO477T, 57 p.

Rogers/Pacific, 1994, Report Accompanying Map of Landslides and Surficial Deposits of the City of Orinda, California for City of Orinda, Department of Public Works: unpublished consultant's report, February 18, 161 p.

Sloan, D., 1992, The Yerba Buena Mud: Record of the last-interglacial predecessor of San Francisco Bay, California: Bulletin of the Geological Society of America, v. 104, n. 6, pp. 716-727.

Regional Versus Site Specific Spatial Hazard Analysis

Matthew A. Mabey[1]

ABSTRACT

The process of hazard analysis, on both a site specific basis as well as a regional basis, relies on the interpretation, interpolation, and extrapolation of sampled data so as to accommodate both the limitations in sampling resolution and the geologic variability present. Arriving at as accurate a depiction of the physical world as possible is the desired result. In analyzing the hazard that the physical world presents we are also faced with limitations and uncertainties in our ability to assess the hazard implications of the materials present. But regardless of the absolute accuracy and precision of our hazard analysis tools, in the end the best possible assessment will result from developing as accurate a model of the physical world as possible.

Several different approaches to dealing with the sampling resolution and geologic variability hazards have been proposed and implemented to produce an assessment of earthquake. To date all have involved assumptions, simplifications, and short comings but still resulted in useful products. Examples of the various types of approaches illustrate these strengths and weakness. Improved techniques suggest themselves by building upon past successes and modern technology.

Combining regional and site specific hazard assessments into risk assessments is a natural application of hazard information. This must be done with care and in consideration of the same spatial resolution and variability issues.

Spatial assessment of hazards has progressed up to the present in parallel with the tools and techniques for borehole specific hazard assessments. Opportunities for improved techniques now present themselves. The spatial assessment technique appropriate for a given application will depend on numerous factors specific to the application

INTRODUCTION

In addressing issues relating to the scale and resolution of spatial hazard

---

[1]Assistant Professor, Brigham Young University, Dept. of Geology, P.O. Box 25111, Provo, UT 84602-5111

assessment, the objective here is to focus on some of the key issues arising from the fact that there is no way of knowing exactly what materials exist everywhere, in any study region. Earthquake hazard assessment is explicitly discussed here, but the concepts are portable to other hazards. No attempt was made to be comprehensive in reviewing past mapping efforts. Rather a few select examples are mentioned to make the issues tangible. The examples were selected because they were judged, by the author, to be high quality implementations of legitimate approaches to hazard mapping that exhibit the effects of the necessary simplifications and assumptions. The figures used are fictitious rather than real examples in order to clearly demonstrate the concepts.

SAMPLING AND VARIABILITY

The first set of issues concern the resolution and distribution of data, or samples, and their relationship to the variability of the geology (lithologies and facies) and the geotechnical properties (shear wave velocity, blow count, plastic limit, liquid limit, void ratio, unit weight, grain size distribution, etc.)

Borehole Sampling and Logging

It is theoretically possible to drill a cluster of three holes immediately next to one another. The first being a continuous core, the second collecting split spoon samples ever two to three feet and the third collecting undisturbed samples in all materials suitable for advancing a shelby tube. This would, in one dimension, give an essentially continuous sampling of information needed to characterize that square meter of land. For practical reasons this is not done and so, what results at a drill site depends on the drilling technique as well as the sampling and testing procedures being used. The information is incomplete, even regarding the materials actually penetrated by the drilling. The levels of lithologic contacts will be approximate. Even it Standard Penetration Tests (SPT) are performed every 5 feet, a layer of five blow material over three feet thick could be completely missed by a very thorough site investigation. (see Figure 1). Additionally most geotechnical investigations are limited to depths needed to accommodate planned excavation and or bearing capacity requirements. Some relatively shallow materials go completely uninvestigated. So in practice even at the site of a borehole, the resolution and extent of the data collected is limited and interpretations, interpolations and, extrapolations are called for.

Borehole Distribution

As the hazard assessment moves away from the actual location of borehole data, both the geology and the geotechnical parameters may change. Are changes present? Are they abrupt or gradual? Are they linear or are the exponential? Definitive answers to such questions come only from drilling more boreholes. Reasonable estimations result from approaches described later but for now the limited number, the pattern of distribution, and the type of subsurface data points is of concern. Figure 2 shows a hypothetical city with a variety of different kinds of

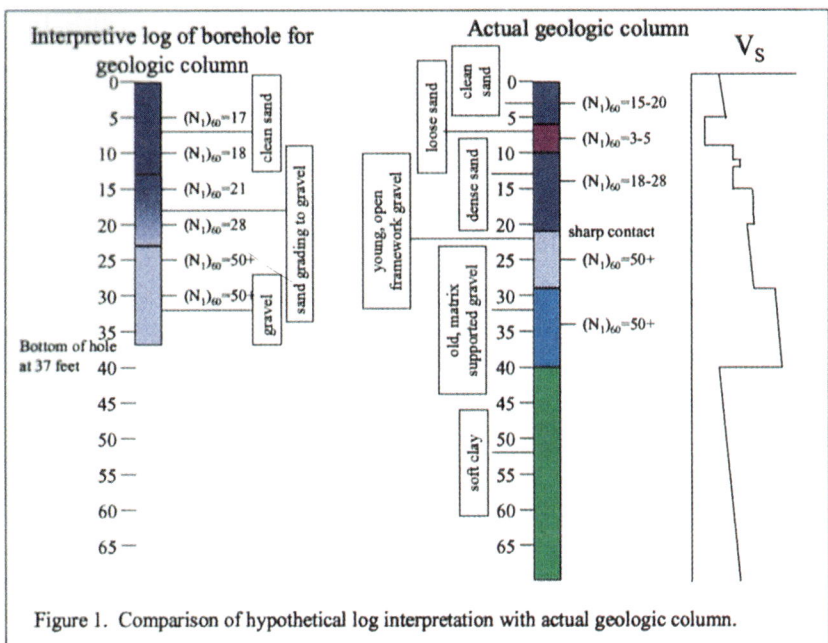

Figure 1. Comparison of hypothetical log interpretation with actual geologic column.

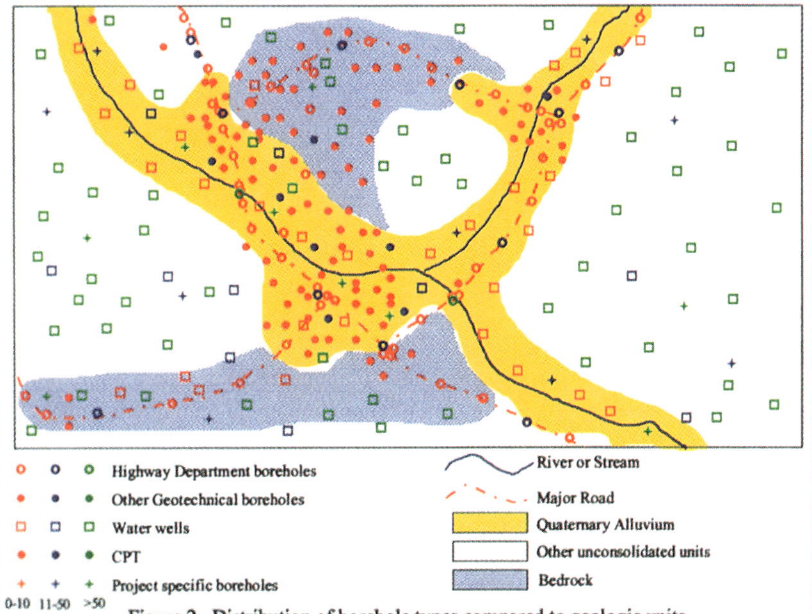

Highway Department boreholes
Other Geotechnical boreholes
Water wells
CPT
Project specific boreholes

0-10 11-50 >50
m   m   m

River or Stream
Major Road
Quaternary Alluvium
Other unconsolidated units
Bedrock

Figure 2. Distribution of borehole types compared to geologic units

subsurface data points and some typical distribution patterns. The symbols represent the type of data point and the color indicates the depth penetrated.

One data type indicated are boreholes drilled along state highway corridors. Doroudan and others (1996) describe a large, high quality database which was assembled for the purpose of seismic microzonation mapping. Their map of the distribution of data points resembles the distribution of highway department boreholes here. The problem is that the data is all concentrated in narrow bands where the highways are, leaving large areas completely unsampled. In addition to the problem of uneven distribution is the fact that roads tend to pass through a limited range of types of topography and/or land use types.

A second class of borehole data consists of geotechnical boreholes. These will typically be as abundant as, or more abundant than, the highway department boreholes. The distribution pattern of these holes will not be limited to narrow highway corridors but will still be uneven and limited to areas where large engineering projects have been completed or contemplated. Large areas are still either unsampled or little sampled and these are the areas where future development will likely take place. These geotechnical boreholes are often to depths of only 10 meters (30 feet) or less.

The next borehole data type is a water well. Water wells tend to be more evenly distributed, including in areas where limited development has occurred to date. They also tend to average a greater depth of penetration. Unfortunately the only information typically collected in water well drilling is lithologic and water information. Often the lithologic information is of poor or unreliable quality. If nothing else major contacts and depth to bedrock are usually recorverable from water well data.

Cone penetrometer data tends to be less abundant than the other types. The depth of penetration is also limited. But the data is very useful for liquefaction analysis and in that it gives a continuous sequence of data regarding the subsurface.

The last type of subsurface data point depicted in Figure 2 are purpose drilled boreholes. This is the most valuable data (on a hole by hole basis) for spatial hazard assessment, but also the most expensive. These purpose drilled boreholes are valuable because they can be located where they are needed, to how deep is needed, of known data quality, and include all the tests and sampling desired (i.e. infrequently collected data such as shear wave velocity). For a given hazard assessment none or all of these data types may be available. But since many are in the public domain and most require only time to collect, as much subsurface data as possible should be collected.

Surface Mapping and Imagery

Another source of information useful for spatial hazard assessment are surface maps (i.e. geologic maps and flood-plain maps), air photos, and satellite imagery. These data may or may not represent a greater data sample density than the boreholes. They have the advantage of giving a continuous depiction of the geology or other useful properties such as vegetative cover. They are also data that is available for most areas at some scale.

The disadvantage of this data is that it is two dimensional, dealing only with what is at the surface. The conditions and materials existing even at a depth of a few feet can vary widely with identical surface conditions.

Geophysical techniques

Geophysical techniques can be employed to good advantage at both site specific and regional scales, particularly if correlations to borehole subsurface data are available. The depths of greatest interest to hazard mapping are typically only well sampled by geophysical techniques specifically applied with near surface conditions in mind.

An example of an effective application of shallow geophysical data is found in Mabey and others (1993) where shallow seismic refraction was used to fill in an area of sparser borehole data to model the thickness of a unit of variable thickness that was critical to slope stability assessment.

Three Dimensional Geology and Geotechnical Property Variability

As pointed out in the preceding sections the geology and geotechnical properties can be sampled to a certain degree of resolution. In one dimension at boreholes and two dimensionally at the surface. However even the logs and maps require a degree of interpretation, interpolation and extrapolation. As a more complete model of how the geology and geotechnical parameters vary three dimensionally this process in continued. In some settings the conditions encountered in the boreholes of a single building site can vary dramatically while at other times two boreholes drilled miles apart will indicate equivalent conditions. The variations can be in the form of the changes in lithology and thickness or changes in a measured properties such as shear wave velocity or clay content. Data sets and sampling can be analyzed to discern whether the variability is perhaps spatially random about some mean and with some distribution or is systematic, perhaps varying from one mean at the north end of a study region or site to a significantly different mean at the south. When working within urban and suburban or site specific areas, enough data is likely available to pursue such patterns. In regions of less existing development, assumptions of certain degrees of uniformity are more likely to be necessary.

ANALYSIS APPROACHES

There are numerous approaches to analyzing geologic conditions to arrive at an assessment of the hazard. Modern computing resources allow for many site specific analysis tools to be applied to regional mapping as well as site specific assessments. This is especially true of the liquefaction and ground shaking hazards (Mabey and others, 1993; Rockaway and others, in press; Sharma and Kovacs, 1982; Levson and others, 1996). Some forms of slope stability analysis can also be applied on a regional scale (McCalpin, 1996; Mabey and others, 1993) but others have yet to be applied due to the detailed data requirements. There are no fundamental impediments to development of more refined slope stability analysis on regional scales in the future.

The key aspect of analysis tools is that the accuracy and reliability of the results for all the techniques is tied to the accuracy and reliability of the input parameters. The effects of generalizations and simplifications required for regional assessment can be compared to results at locations where detailed information is available. Confidence in the simplifications and generalizations can be gained from this. At all scales, deviations of interpretations from reality between data samples and limitations of the analysis tools always have the potential to introduce inaccuracies. Thomas and McFadden (1995) found variations in both sampling accuracy and geology that would be completely overlooked in even a thorough, but standard, site investigation.

Once the hazard has been systematically and objectively quantified, then the hazard can be simplified into qualitative categories for specific applications. The best possible assessment at any scale will result from using as much data as possible, with great of care and skill as possible and the best tools possible. If this is done, reasonable and useful products will result at any scale.

EXTRAPOLATION AND INTERPOLATION APPROACHES

On a site specific scale cross-sections through subsurface data points are constructed based on geologic assumptions of the nature of the transitions between points (gradual versus abrupt) and then design parameters are calculated based on these assumptions and the resulting model. This process is little questioned due to the relatively small distances over which the interpretations, interpolations and extrapolations are made. This so, not withstanding the fact that very dramatic geologic and geotechnical changes can easily occur at such scales. When the interpretations, interpolations and extrapolations are done on a more regional scale it is more common to question the process even though the same caveats apply. The following are six approaches to this process that have been proposed and, with one exception, usefully applied at regional scales. The approaches also have validity for application at a site specific scale for more objective results.

Interpreted Geologic Map

In general certain ages and types of geologic materials have certain characteristics regarding the hazard they result in. This is largely related to the degree of consolidation and the apparent plasticity of the units. For example Youd and Perkins (1978) proposed an age and depositional environment based scheme for classifying the liquefaction susceptibility of geologic units. Madin (1990) developed a set of simple hazard maps for a part of the Portland area based on this type of simple interpretation of the geologic units. The maps also included contours of the thickness of quaternary sediments and depth to "bedrock". These maps were applied to a number of uses. In retrospect they also had some flaws.

One was that by saying a given unit is susceptible to liquefaction users of the map tend to get an impression of too severe a liquefaction hazard because users tend to over look the fact that the water table is not everywhere near the surface where the unit is present. This is inspite of statements to that affect in the text of the map. Another short coming is that in the absence of actual geotechnical data and analysis many people are disposed to challenge the information as overstating the hazard (often equally in the absence of data and analysis). The reverse of this problem is that hazards are sometime surprising and defy conventional wisdom, for example a Plio-Pleistocene silt unit that is still liquefiable as indicated by both analysis and evidence of Holocene liquefaction. None-the-less a broad but reasonable portrayal of the hazard is possible based simply on a thoughtful interpretation of a geologic map (see Figure 3).

Evaluation of Geologic Map Units

A somewhat more sophisticated approach is to perform analyses on a limited number of sample points in order to directly assess the hazard implications of local geologic units. This can often be done with data from a single one of the sources of subsurface data listed earlier or using the files of a single geotechnical consultant. This was the approach used by Grant and others (1981) to construct a liquefaction hazard map of the Seattle, Washington area. This approach is very common and gives results whose accuracy is always improved by increasing the number of sample points. Imagery can also be a part of this process.

The first short coming of this approach is that it ignores the third dimension and assumes that what is at the surface extends to a great enough depth that what is beneath is unimportant to the hazard. Obviously this is not always the case. Another major short coming of a view of the world which is centered on mapped geologic units is that systematic variations of geotechnical properties within a unit may be overlooked. This was found in Seattle when Mabey and Youd analyzed the variation of SPT blow counts as a function of positions in the Duwamish Valley (edge versus center). The result being an over estimate of the hazard at the edge of the valley (see Figure 4).

Generalization of Hazard Into Large Cells or Polygons

Another scheme which can be motivated by either a desire to limit the computational effort or by an attempt to match subdivisions of a different data set (i.e. demographic) is to assess or generalize the hazard into large (kilometer dimension) regular cells or into polygons such as census tracts or zip code zones.

The "HAZUS" damage and loss estimation methodology (Whitman and others, 1996) employs this approach for analyzing the general building stock and population. In general this is effective but on the periphery of urban areas the size of these polygons becomes overly large and the variation of geotechnical hazards within a polygon can become overly great. Also a problem is the variation of the construction and population with in the polygon (see figure 5). King and Kiremidjian (1996) modified this approach with a process of subdividing

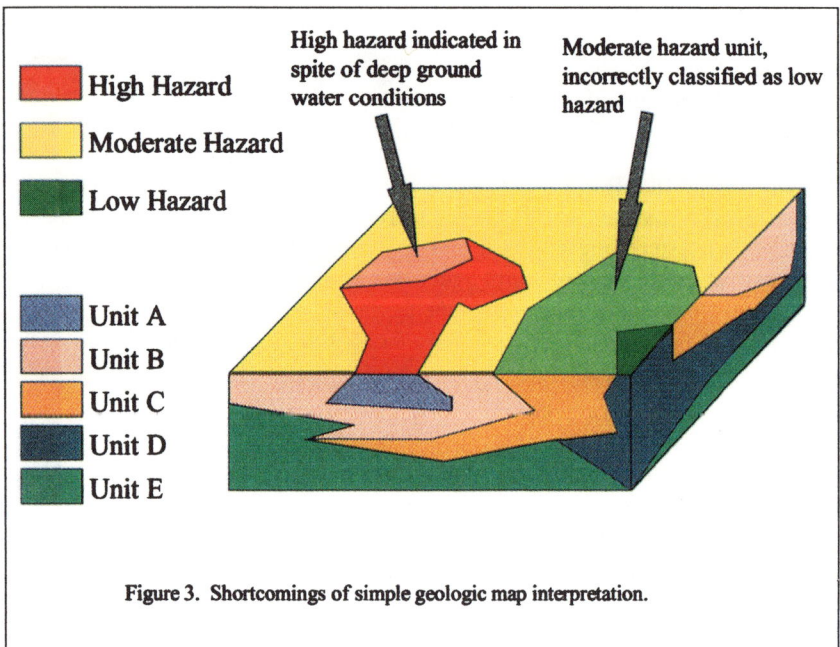

High Hazard

Moderate Hazard

Low Hazard

Unit A
Unit B
Unit C
Unit D
Unit E

High hazard indicated in spite of deep ground water conditions

Moderate hazard unit, incorrectly classified as low hazard

Figure 3. Shortcomings of simple geologic map interpretation.

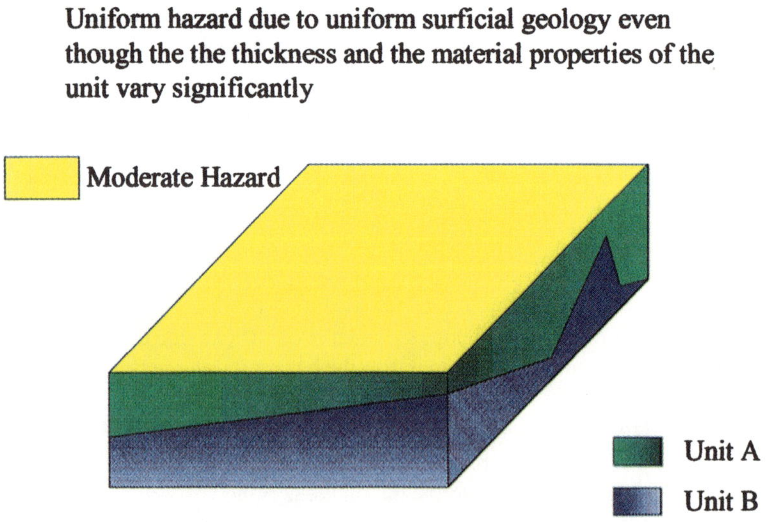

Uniform hazard due to uniform surficial geology even though the the thickness and the material properties of the unit vary significantly

Moderate Hazard

Unit A
Unit B

Figure 4. Using analysis of geologic units.

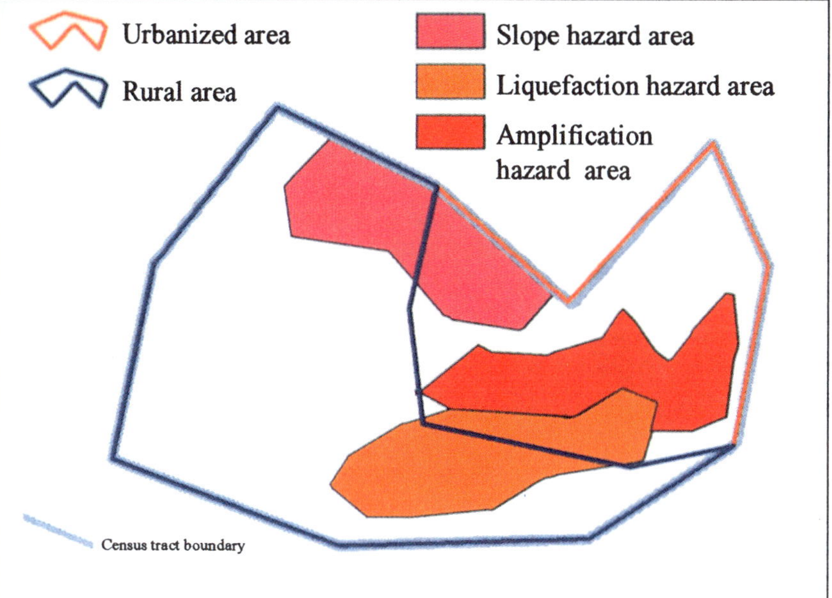

Figure 5. Pitfalls of using large areas such as census tracts

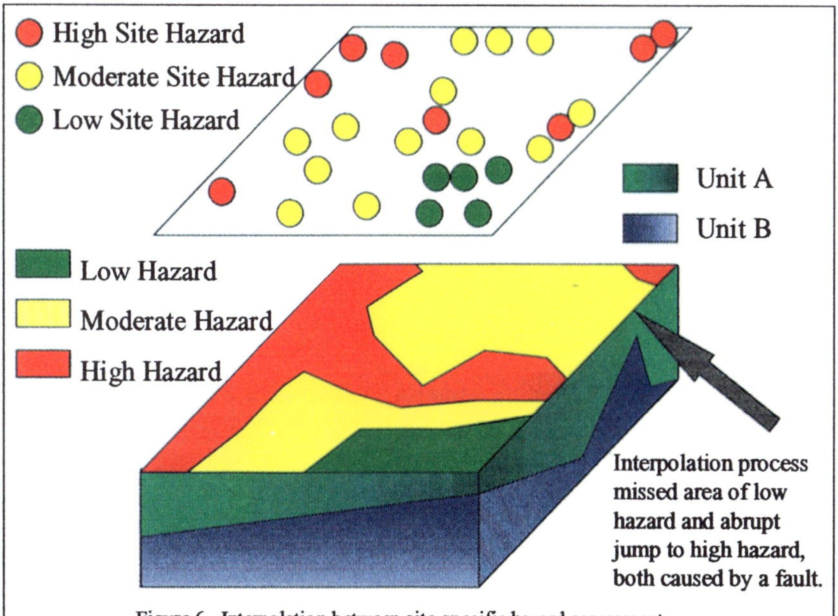

Figure 6. Interpolation between site-specific hazard assessments

demographic units to accommodate changes in qualitative catagories of hazard. Care must be take to be aware of this problem and its implications.

Interpolation of Hazard Between Data Points

Another approach that has been employed is to assess the hazard on a site specific basis at a (sometimes large) number of data points and then interpolate the hazard between data points rather than the geology and geotechnical properties. This approach was used by both Sharma and Kovacs (1982) and Rockaway and others (in press). This assures that the assessment of hazard at the location of data points is as accurate as possible and then relies on the interpolations between these points to achieve reasonably close approximations of the conditions in between.

The difficulty with this approach is that strictly machine based interpolation schemes will ignore geologic conditions, boundaries, and principles. Likewise interpolations done by hand may ignore or inadequately accommodate geologic conditions. The machine interpolations can be guided by constraints and "false data" to be more geologically sound but often the connection between geology, geotechnical parameters, and the resulting hazard is not clear and intuitive (see figure 6).

Interpolation of Three Dimensional Geology Between Data Points

A refinement of the evaluation of geologic map units is to extend the geologic mapping, or model, into the third dimension and explicitly include the thickness of not only the surficial unit but also units beneath the surficial unit. An example of this approach is the Relative Earthquake Hazard Mapping Project for the Portland area (Mabey and others 1993). This is done by collecting data from all possible sources to have the maximum possible resolution, both horizontally and vertically. Project specific data is an important compliment to this approach. The hazard analysis can then be applied in a site specific manner to site-specific-like geologic column for a grid of cells with dimensions of tens of meters (see Figure 7).

The short coming to this approach, as applied to date has been the use of average geotechnical properties for a given unit in the site-specific-like geologic column. This allows the hazard assessment to deviate from a true site-specific assessment's results at the locations of data points. In the application in Portland the variability of the unit properties was such that the difference between the true site specific analyses and the grid cell analyses was not significant. This would not be true in all geologic settings.

Independent Interpolation of Both Three Dimensional Geology and Geotechnical Properties Between Data Points

A combination of the last two techniques suggests itself. The geologic model could be constructed in three dimensions as outlined in Mabey and others (1993). But instead of using averaged geotechnical parameters, the geotechnical parameters used in a given cell could be interpolated in a mostly automatic fashion as in Rockway and others (in press) from the actual data points. This would allow

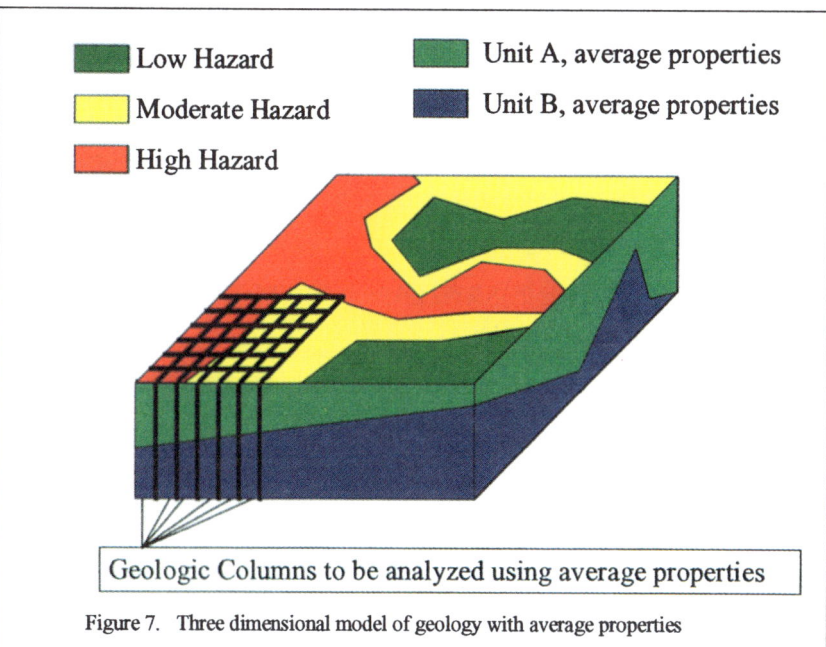

Low Hazard   Unit A, average properties

Moderate Hazard   Unit B, average properties

High Hazard

Geologic Columns to be analyzed using average properties

Figure 7. Three dimensional model of geology with average properties

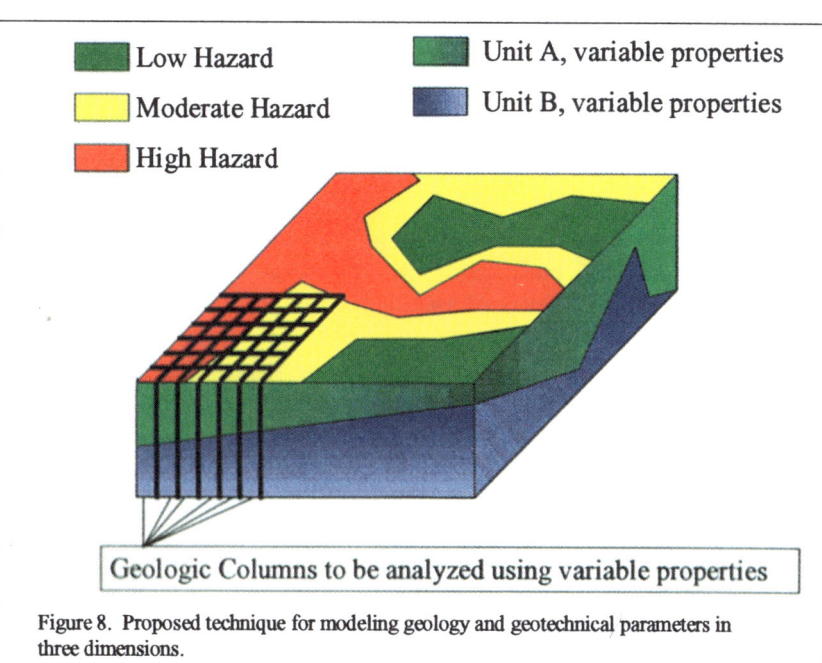

Low Hazard   Unit A, variable properties

Moderate Hazard   Unit B, variable properties

High Hazard

Geologic Columns to be analyzed using variable properties

Figure 8. Proposed technique for modeling geology and geotechnical parameters in three dimensions.

systematic, as opposed to random, variations in geotechnical parameters as well as the best possible interpretation of the geology to be included in the hazard analysis (see Figure 8). This approach combined with strategically positioned, project specific drilling and some geophysical techniques should result in an accurate and reliable depiction of the hazard. Current technology is both capable of, and affordable for, moving on to this next step in refinement.

APPLICATIONS TO RISK ANALYSIS

Hazard assessments have a multitude of uses but one of the key uses in recent years has been their utilization in risk assessment (King and Kiremidjian, 1996; McCormack, 1996). The limitations of the hazards assessment approaches discussed above must be carefully considered and accommodated when using them as part of a risk assessment.

Likewise the principles and pitfalls which apply to the hazard assessments also apply to the risk assessment process. Most risk assessments involve a process where the distribution of building and land-use types, as well as population, are assumed to be uniform across large polygonal areas such as census tracts and zip code zones. While this assumption is sometimes reasonable, there are other examples where it is definitely not. Likewise the practice of assigning uniform hazard across these same large polygonal areas can be a serious mistake that will result in either a dramatic overestimate or underestimate of the risk (see Figure 5).

Census tract and zip code zone boundaries are an artifact of data set sampling resolution and have no intrinsic relationship to either hazard or risk and therefore should not be preserved at the expense of degrading the resolution and accuracy of other data sets.

CONCLUSION

Spatial assessment of hazards has progressed up to the present state of the art in parallel with the tools and techniques for site-specific borehole hazard assessments. Reasonable interpretations of the hazard between data points are possible. Opportunities for improved techniques now present themselves. The spatial assessment technique appropriate for a given application will depend on numerous factors specific to the application. These factors include both the resolution and quality of available data as well as the geologic setting.

REFERENCES

Doroudan, M., Vucetic, M., and Martin, G.R. (1996). "Development of 3-dimensional geotechnical database for Los Angeles seismic microzonation." *in*: Proceedings of the Eleventh World Conference on Earthquake Engineering, Acapulco, Mexico, Elsevier Science Inc., Tarrytown, New York, 1454-1461.

Grant, W.P., Perkins, W.J., and Youd, T.L. (1991). Evaluation of liquefaction potential, Seattle, Washington. U.S. Geological Survey Open-File Report 91-441, 44p.

King, S.A. and Kiremidjian, A.S. (1996). "Shake, rattle, and map." Civil Engineering, ASCE, New York City, June, 1996, 50-52.

Levson, V.M., Monahan, P.A., Meldrum, D.G., Matysek, P.F., Gerath, R.F., Watts, B.D., Sy, A., and Yan, L. (1996). "Surficial geology and earthquake hazard mapping, Chilliwack, British Columbia, *in*: Geological Fieldwork 1995, Grant B.M. and Newell, J.M., eds., British Columbia Geological Survey Paper 1996-1, 191-203.

Mabey, M.A., Madin, I.P., Youd, T.L. and Jones, C.F. (1993). Earthquake hazard maps of the Portland Quadrangle, Multnomah and Washington Counties, Oregon, and Clark County, Washington, Oregon. Department of Geology and Mineral Industries, Portland, Oregon, GMS-79.

Mabey, M.A. and Youd, T.L. (1992) Prediction of displacements due to liquefaction induced lateral spreading, Brigham Young University Department of Civil Engineering Report No. CEG 92-02.

Madin, I.P. (1990). Earthquake-hazard geology maps of the Portland metropolitan area, Oregon—text and map explanation. Oregon Department of Geology and Mineral Industries Open-File Report 90-2, scale 1:24:000, 21p.

McCalpin, J.P. (1996). "Earthquake-induced landsliding analyzed and predicted with a geographic information system." *in*: *Natural Disaster Reduction*, Housner, G.W. and Chung, R.M., eds., ASCE, New York City, 3-4.

McCormack, T.C. (1996). "A Methodology for Regional Seismic Damage Assessment and Rehabilitation of Existing Buildings." Ph.D. Dissertation, Portland State University, Portland, Oregon.

Rockaway, T.D., Frost, J.D., Eggert, D.L., and Luna, R. (in press). "Spatial earthquake hazard assessment of Evansville, Indiana*." Seismological Research Letters*, Special Volume: Structural Framework and Tectonic History of the Illinois Basin/Wabash Valley.

Whitman, R.V., Lagorio, H.J., and Schneider, P.J. (1996). "FEMA-NIBS earthquake loss estimation methodology." ." *in*: *Natural Disaster Reduction*, Housner, G.W. and Chung, R.M., eds., ASCE, New York City, 113-114.

Sharma, S. and Kovacs, W.D. (1982). "Preliminary microzonations of the Memphis, Tennessee area." *Bulletin of the Seismological Society of America*, vol. 72, no. 3, 1011-1024.

Thomas, T.N. and McFadden, B. (1995). "Results of the liquefaction test site study." May 6, 1995 Seminar on In Situ Testing for Seismic Evaluation, ASCE Seattle, Washington Geotechnical Section.

Youd, T.L., and Perkins, D.M. (1978). "Mapping liquefaction-induced ground failure potential." Journal of the Geotechnical Engineering Division, ASCE, 104(GT4), 433-446.

Youd, T.L. and Perkins, J.B. (1987). Map showing liquefaction susceptibility of San Mateo County, California. USGS Miscellaneous Investigations Series, Map I-1257-G, scale 1:62,500.

Spatial Data Quality Evaluation in Geotechnical Earthquake Engineering

Ronaldo Luna [1]

Abstract

Methods of analysis and processing of spatial data have become prevalent in engineering and have recently been a source of study in the evaluation of earthquake hazards. Numerical routines in GIS and other spatial information systems use databases that often reside away from the easy access of the engineer and may consist of variable data quality levels. Once a spatial database is implemented it is rarely scrutinized for its correctness, completeness or accuracy, leaving behind serious data quality issues. Since data quality is relative to the application where the data is used, this paper focuses on the issues related to spatial geotechnical hazard evaluation. General topics of spatial information systems are reviewed before a more detailed evaluation of the pertinent geotechnical applications. An example of the evaluation of boreholes logs used in a spatial liquefaction evaluation is also presented.

Introduction

Computer-based tools used to work with data that evaluate phenomena located on, above or below the earth's surface have undergone major recent advances. These tools have evolved to become systems that utilize data of many forms and continuously undergo improvements and technological advances. A geographic information system (GIS) is a type of spatial information system and the term "spatial" is used here to refer to located data or objects positioned in any space, not just geographical. These systems are based on a collection of data stored typically in a databank or database that can be represented in the form of a map. The expression "the data drives the map" alludes to the fact the data is the foundation for a good final result with spatial reference, often geographical. A lot of time and energy has been devoted to developing tools and methods that can manipulate, analyze and display data, however, little attention has been paid to the quality of the data or a means to evaluate it.

---

[1] Assistant Professor, Department of Civil and Environmental Engineering, Tulane University, New Orleans, LA 70118.

Data, information and knowledge may seem like very different terms, however, in the context of a database these terms become much more interrelated. Information is derived from individual or a combination of data elements in a database - information that is not directly apparent. In a sense, the information is produced from data by use of our thought process, intuition, judgment based on our knowledge. The development of a spatial geotechnical database consists of a combination of data from different investigators in one area/region conducted at different times and it is considered that by using available information the study becomes more comprehensive and our knowledge about the site is improved. In this information age, a database that provides information and affects our knowledge is often considered a commodity, and data quality often marks a difference in value.

Data quality is essential for the adequate use in any application. Quality has numerous definitions in different fields of science and applied science, but one rather accepted definition is "fitness for use". This definition enables users to make a judgment for their particular application. It should be apparent to the user that no representation is a perfect replica of something as complex as the earth. Quality has different aspects that are of importance to users and can be used to determine the fitness of the data for their particular use (Aalders, et al., 1995). This paper presents the important issues related to data quality as it pertains to spatial information systems and specifically the applications in geotechnical earthquake engineering. A method for data quality evaluation as it relates to spatial liquefaction analysis is discussed in more detail. This method is further explained in the context of a geotechnical database for Treasure Island, California designed for use with a GIS that was subjected to data quality evaluation.

Data Quality in Spatial Information Systems

Data quality is important for any application using a spatial database and there are data quality factors that should be considered. Aronoff (1989) summarized the characteristics that affect the usefulness of data as it relates to data quality and divided them in three categories: micro level, macro level and usage, as shown in Table 1. These categories are further divided into components with their respective description. The data quality characteristics outlined in Table 1 were collected for the general purpose of fitness for use in a spatial information system and are only applicable for this purpose. A more elaborate and detailed list of characteristics would be required for a particular application.

The following is an example related to data quality in the development of a spatial coverage. Say an existing map of scale 1:100,000 is used to digitize a polygon representing the boundaries of a site, then this polygon is moved and used on another coverage originally developed from a USGS quad sheet scale 1:24,000. The process of going from one spatial coverage to another is a very common operation, however, going from a small scale (large area) to a large scale (small area) is something that will decrease the quality of the data produced. This example crosses several data quality components, such as, positional accuracy, resolution, lineage, scale and source of data.

To manage data quality in a spatial information systems one needs to understand the type of errors involved when handling spatial data. There is always a degree of error in spatial information and error is introduced at every step in the process of generating and using spatial data sets, from collection of the source data to the interpretation of the results in a completed analysis. However, the objective in dealing with error should

Table 1 - Spatial Data Quality Characteristics

Category	Components	Description
Micro Level	Positional Accuracy	the expected deviance in the geographic location of an object in the dataset from its true ground position.
	Attribute Accuracy	the accuracy of discrete or continuous variables serving as attributes.
	Logical Consistency	refers to how well logical relations among data elements are maintained.
	Resolution	the smallest discernible unit or the smallest unit represented.
Macro Level	Completeness	aereal and contents coverage, proper classification used, and amount and distribution of field measurements.
	Time	the temporal dependence of data and how it affects its significance.
	Lineage (Legacy)	the dataset history, the source and processing steps used to produce it.
Usage	Accessibility	the ease of obtaining and using data; data restrictions and public/proprietary data.
	Direct/Indirect Costs	the direct price paid for the data; the indirect cost including the time and materials used to create the data.

not be to eliminate it, but to manage it (Aronoff, 1989). *Data collection* errors are common because errors often exist in the original source materials that are entered into the spatial database. These errors may be a result of inaccuracies in field measurements, inaccurate equipment, or incorrect recording procedures. *Data input* errors are those due to positional errors, for example, digitizing tablets are commonly accurate to fractions of a millimeter but it varies over the digitizing surface (more accurate in the center). *Data storage* in digital form requires a finite level of precision and the use of appropriate significant digits. Today we can find computers that store data in different memory size (8, 16, 32-bit), however, you may overcome this problem by the use of extended (double) precision. *Data manipulation* may involve procedures such as combining multiple overlays and as the number of overlays increase, the number of possible opportunities for error increase. Also, the development of thematic maps in a raster format file may not be detailed enough to represent the area (i.e., size of cell too large). *Data output* errors can be introduced in the plotting of maps due to the plot/print device or the media used to print on which may shrink or stretch. *Use of results* is something difficulty to control since the user may not be affiliated in any way to the author of the data. Reports or maps may be misinterpreted, accuracy levels ignored, and inappropriate analyses accepted. This last source of error may seem independent to the system, but in fact, the resulting errors in decision making represent errors in the process spatial data usage.

Data Sources and Requirements

For the development of an engineering database the formal process of defining the data sources and data requirements is an important task that may lead to improved data quality. Even though geotechnical data typically encompasses subsurface information, other forms of data (e.g., roads, rivers, site boundaries, etc.) are required to complement the geotechnical data and make it useful in a spatial framework. Table 2 shows a partial list of different data sources for a geotechnical project classified by their category and type.

Table 2 - List of Typical Geotechnical (or related) Data Sources

Data Category	Data Source or Type	Content of Data
Geotechnical Subsurface	Geotechnical Reports	Subsurface conditions as they relate to foundation engineering and soil mechanics containing specific engineering recommendations.
	Geologic Reports	Regional geology, geomorphology, or structural geology documented for general usage.
	Boring (Borehole) Logs	Subsurface conditions, stratigraphy, drilling, sampling, field and laboratory testing, etc.
	CPT Logs	Tip resistance, side friction and pore pressure during penetration. Permeability.
	Test Pit Logs	Excavation properties, stratigraphy, and soil descriptions.
	Piezometers and Groundwater Wells Logs	Groundwater changes and pore water pressures
Geophysical Subsurface	Geophysical Borehole	cross-hole testing data and continuous
	Geophysical Reflection and Refraction	Stratigraphy, dynamic soil properties.
	Ground Penetrating Radar	Stratigraphy of subsurface and subsurface anomalies
Seismic	Earthquake Events	Location, hypocenter, epicenter, magnitude, and acceleration of event. (NCEER).
	Seismic Maps	Locations of faults, background sources and other earthquake source.
Infrastructure and	Regional DLG and DEM	Spatial USGS data including transportation, hypsography, hydrological, political boundaries, etc.
Demographic	TIGER Data	Spatial census information with demographics and regional features
	CADD Drawings	Roads, buildings, utilities
Images	High and Low Altitude Aerial Photographs	Image of surficial features at a specific time (buildings, roads, EQ hazards and damage areas)
	Photographs	General interest site photographs to aid in the reconossaince. (damage areas and documentation of hazards may be recorded)

45

Once the data sources have been identified independent of a particular project, one should identify what is required to solve the problem statement. This can be as simple as listing all the requirements and organizing them as a function of their relationships. Geotechnical data also depends on other data and may require corrections of the measured value as a function of depth. For example, the resistance to penetration SPT N-value is dependent on information available in the geotechnical report, such as, drilling, sampling and testing methods. Relationships between these different data types can be established in a database once it is logically organized and structured. This task is usually accomplished early on the project and may have a major impact on the design and implementation of the database. This level of organization and entity relationships start defining the data structure of the database and may affect the use and quality of the data. Most data sources listed in Table 2 have been available for some time now and are used on a routine basis in engineering. This common repeated use of data type usually results on a particular level of standardization.

Data Standards

Data becomes standardized when several users can perform operations on the data set and this may lead to the issue of data transfer. The format and structure of spatial data used in computer systems have become "standardized" for some data types, such as CADD drawings, digital line graphs (DLG) and digital elevation models (DEM). Historically, this standardization has occurred in an ad-hoc manner resulting in defacto standards. For example, one of the most popular data formats is the direct exchange format (DXF) originally developed for CADD drawings. This data format does not support topology, extensive attribute data, and geographic coordinates, which are fundamental for a GIS. However, this has not limited other software developers and data producers creating new data formats. The issues of spatial data standardization have been addressed by national commissions such as, the Federal Geographic Data Committee (FGDC), the National Spatial Data Infrastructure (NSDI), and more recently the National Geospatial Data Clearinghouse (NGDC). The Spatial Data Transfer Standard (SDTS) has been tested, modified and refined to become the NIST approved Federal Information Processing Standard (FIPS) 173 (GeoInfo Systems, 1994; Sarasua and Nissalke, 1993).

A spatial data standard important to data transfer is the "Content Standards for Digital Geospatial Metadata" developed by the FGDC. It was developed to specify the content of metadata for a set of digital geospatial data and provide a common set of terminology and definitions related to metadata. The final version of this data transfer standard was adopted in June 1994. Metadata is, in effect, data about data. Metadata includes data about the content, quality, condition and other characteristics of digital data. In geotechnical engineering borehole logs, metadata is provide at the header information of the document or key (legend) to logs. All the information in the header conditions the data provided in the body (depth related data) of the log, however, there are no specific standards as to the content, quality, or completeness of this geotechnical metadata.

The American Society for Testing and Materials (ASTM) has developed standard methods for testing geomaterials used in geotechnical and geoenvironmental engineering. There are no formal data standards for digital geotechnical data structure and transfer, however, there are some recent initiatives at a national level for this purpose. The National Geotechnical Experimentation Sites (NGES) together with the Federal Highway Administration (FHWA) have published a data dictionary for use in the collection of geotechnical data which will facilitate data compatibility and transfer (NGES, 1993). This proposed data dictionary was based on the British Association of Geotechnical Specialists (AGS) which has been implemented in the UK (Threadgold, 1992).

Quality Issues in Geotechnical Earthquake Engineering

The specific quality issues for geotechnical data need to be addressed by the engineers that measure, collect and use the data. Venkataramanan (1996) conducted a review of the existing data quality studies in geotechnical engineering and found that this topic has not been addressed much in the literature. Limited studies were available such as borehole log data quality by D'Andria, et al. (1995), deep foundations database by Long (1994), and knowledge-based systems to interpret geotechnical information by Troll (1994). Data quality issues start at the source where data is created or collected, for which we need to refer to Table 2. In geotechnical engineering it starts in the field during the subsurface investigation, or even before, when the investigation is being planned and specified. ASTM standards are available for most of the field testing and sampling methods, but they are not always required in practice which makes the standardization and quality of geotechnical data very variable. The advantage of combining the subsurface information with other information in a computer system (spatial database) allows for means of complementing the data where gaps are found. For example, ground water level information may not be available in some borehole logs, but it may be available at nearby locations. If this is the case, interpolation based on ground water level data available, can be used to infer the missing data at the borehole in question. This approach was successfully used by Rockaway, et al. (1997) in a liquefaction study for the Evansville, Indiana area.

Liquefaction

In an attempt to define what are the requirements for data quality for a particular purpose in geotechnical earthquake engineering, an exercise focused on this specific issue was conducted (D'Andria et al., 1995, Luna, 1995). The purpose was to identify the factors that contribute to the quality of geotechnical borings for use in liquefaction evaluation. A technique called "mindmapping" (Buzan, 1976; Wycoff, 1991) was used to conduct a session including a group of researchers and professional engineers to identify factors that should be considered in evaluating the quality of a boring log for use in a liquefaction study. As a result of that session, Figure 1 was created to graphically show the classification and degree of detail involved in the process.

Of course, the items identified in the mindmapping session were comprehensive but not an exhaustive list of requirements and somewhat idealistic in nature. Borings completed for the purpose of liquefaction evaluation usually include only a partial list of the items identified depending on the level of effort spent on the subsurface investigation. In order to define the level of quality a criteria was established. This was done by assigning each one of the items a weight factor depending on how much it was considered to contribute to the overall quality of the data. Once the weights were assigned, a sum of all the weights gave a score that was later ranked to differentiate the borings based on their quality. This method will be presented as an example of data quality evaluation later in the paper and additional details can be found in D'Andria et al. (1995).

47

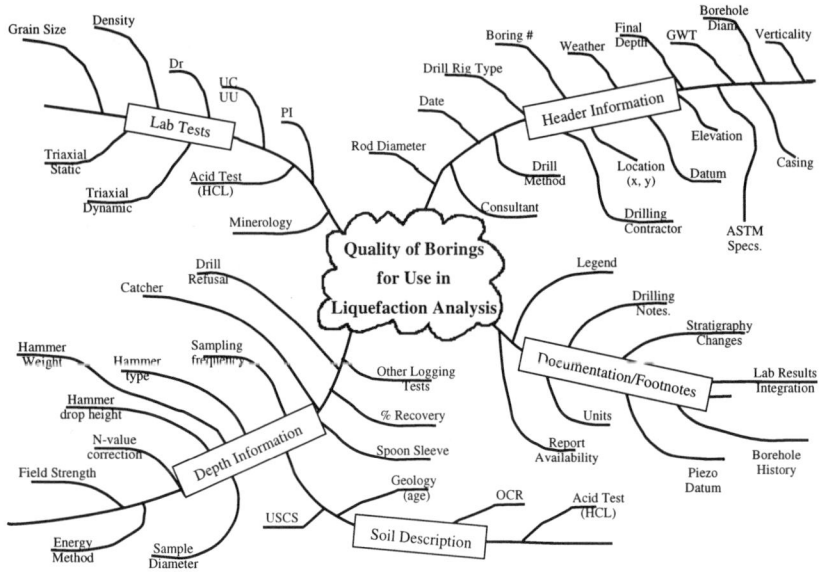

Figure 1 - Mindmapping on Quality of Borings (borehole logs) for Liquefaction.

Ground Motion Amplification (GMA)

The data required for the analysis of ground motion amplification is different than for liquefaction and the emphasis turns to different data types. Evaluating ground motion amplification using a SHAKE type analysis requires the following input data: shear wave velocity (Vs), shear modulus, damping, soil statigraphy, depth to bedrock, and an input ground motion. Shear wave velocities are less commonly obtained in traditional subsurface investigations (e.g., downhole, crosshole, refraction), making the spatial coverage of this dynamic property more sparse. Correlations from corrected N-value are available to estimate shear wave velocities (e.g., Ohta and Goto, 1978; Jamiolkowski, et al., 1988) and can be used to complement the data. However, these two different ways to obtain this dynamic property are clearly two different levels of quality and should be accounted for in data quality evaluation. Another approach to generate data when at a desired location the data is not available, is to use a default dataset (Vs profile) common for the region. This last approach could be classified as a third level of quality and since in engineering this type of inferences are commonly used, they should be addressed accordingly.

In a GMA analysis the depth-to-bedrock (or elevation) is necessary to input the earthquake ground motion. However, in practice not all boreholes are advanced to reach bedrock for economical and practical reasons. This leaves the need to infer the depth-to-bedrock at locations where it was not measured. A powerful inferential technique which is based on spatial statistics or geostatistics is called "kriging". This technique has the ability to model the interpolation parameters (i.e., range, sill, variance) via a variogram and the resolution of the grid in the coverage; additionally, the estimated error can be computed to identify areas of relatively high magnitude of error. Even though this interpolation technique may be considered somewhat sophisticated, it will never be as good as directly measured data, therefore, a lower quality is associated with that analytical approach.

Another important issue is the input ground motion used in the analysis. The earthquake time history may be an earthquake record scaled to the desired magnitude or a synthetic record generated for an area where records are uncommon. Earthquake records scaled to different magnitudes are commonly used in the west coast, however, in the midwest earthquake records are less common and the frequency content is quite different than the west coast records. Martin and Pond (1996) and Rockaway, et al. (1997) have used synthetic records as input motions which can be classified as a different level of quality when compared to the available scaled earthquake records. The issue now becomes defining which approach will provide a higher degree of quality in the analysis.

Earthquake Induced Landslides

This earthquake hazard has been, by far, the one presenting more challenges in geotechnical spatial hazard analysis. To evaluate the potential for ground failure in the form of a landslide, comprehensive data is necessary in the following topics: geometry of the slope, shear strength along postulated failure plane, ground water conditions, and earthquake loading. One of the problems is related to how to tie the subsurface data with the slope that has been identified for analysis. The easy answer is to take the closest borehole log, but we need to specify what distance is acceptable and the borehole log may or may not have the information desired (e.g., shear strength). Obviously, these questions and barriers associated with this data can be categorized with different levels of quality.

The slope of the ground surface to obtain the geometry of the problem can be obtained in different ways. If a DEM is available, the slope at the centroid of the DEM cell can be used to define the sloping ground surface or an average slope may simplify the problem. If a contour digital map is available, a simple cross-section may do a better job depending on the contour intervals and the scale of the source data. Again, different degree of quality can be assigned to the different methods used.

Geotechnical Subsurface Data Quality

Historically the most common means of recording subsurface is by the use of the borehole log or boring log. This document contains data from field and laboratory testing making the information a compilation of measured and interpreted data. Usually the document contains some form of header information that pertains to the general borehole and the depth information that contains sampling, testing results and drilling notes. Evaluating the quality of subsurface data may be a very subjective process and will vary based on the purpose of the evaluation. In an attempt to quantify the data quality of boreholes in a systematic and flexible approach, Venkataramanan (1996) developed a program that aids in this process. The system for data quality analysis and review (QUASAR) implemented the use of the following expression:

$$Q = \frac{\sum_{i=1}^{n} I(i) \cdot R(i)}{\sum_{i=1}^{n} I(i)}$$

where, I is the relative importance of an attribute being evaluated, R is a reduction factor for a second and more detailed level of quality attribute, and Q is the quality parameter that will vary from 0.0 to 1.0. This program has the flexibility of defining different data quality models based on the opinion of experts in the field of application. Once the data quality model is developed, a series of boreholes can be systematically evaluated for their quality and a relative quality can be quantified for each borehole log. A future enhancement to this data quality evaluation process could be to introduce heuristics in the definition of the data model by the use of a knowledge-based system.

Data Quality Evaluation for Liquefaction - An Example

One of the advantages of creating a database for use in a spatial environment (i.e., GIS) is the ability to combine multiple sources of borehole log information. The process of creating a database requires initial data collection efforts for the area of interest. However, a subsurface investigation is typically performed by a consultant/investigator for a specific purpose, and due to budget constraints, only data pertinent to the intended purpose is collected. For example, a borehole advanced as part of a study to determine the shear strength of a soil will use different sampling and testing techniques than a borehole which is drilled to determine the contamination level of the soil profile.

As indicated earlier, a methodology was developed to evaluate the quality of boring logs from a liquefaction analysis perspective. For that purpose and based on the mindmapping exercise presented in Figure 1, the type of information contained in an ideal borehole log was classified in four (4) categories corresponding to the main tree branches in Figure 1 (header information, depth information, lab data and documentation and footnotes). Each of these categories were further divided into subcategories that represent the data items useful for liquefaction analysis. Some of these categorized data items may be rarely found in actual borehole logs, however, they have been included since the purpose of a borehole log quality evaluation method is not to reflect current practice, but to have a reference point of superior quality. The following paragraphs discuss the significance of some selected data items within the framework of liquefaction analysis.

It is well known that the equipment and test methods used to obtain a standard penetration test N-value have an effect on liquefaction analysis. The common procedure to evaluate liquefaction potential in practice can be summarized in the following simple steps: (a) perform boring and sampling using SPT, (b) synthesize results of the field testing (when available) and correct data as appropriate, and (c) calculate the factor of safety for liquefaction using the corrected field data. There are components within each of these steps that may have a significant impact on the computed liquefaction potential (in the form of factor of safety, cyclic stress ratio, or index). During field sampling and borehole logging, factors related to equipment and methods can affect the results. For example, the current state-of-the-practice still adheres to 2.5- or 4-inch diameter borings, however, some practices may use larger diameter boreholes. The effects of testing from relatively large boreholes in cohesive soils is probably negligible, but in sands there are indications that lower N-values may result (Skempton, 1986). Table 3 summarizes the principal SPT equipment components and their corresponding attributes. A relative degree of preference is included in parenthesis.

Table 3 - Common Equipment and Method that Affect the N-value Quality
(adapted from D'Andria, et al., 1995).

Type of Sampler	split spoon (1), California sampler (4), Shelby tube (4). (Riggs, 1986)
Hammer Type weight: drop height:	automatic (1), manual (2), safety (3), donut (3), pinweight (4), downhole (4). 140 lbs (1), anything else (4). 30 inches (1), anything else (4).
Spoon Sleeve Liner	ASTM 1586 does not require to use a liner. Depending on the type of soil, liner might effect the N-Value.

Note: (1) most desirable, (2) good, (3) average, (4) least desirable. Weight factors follow ASTM and standard geotechnical practice.

The quality of a boring log is not only dependent on the equipment and methods used in the field, but also on how the data is represented on the boring log. The depth to ground water table (DGWT) is one of the most important items in a borehole log used for liquefaction evaluation and therefore is presented in more detail as an example. If information pertaining to the DGWT measurements is not recorded, then the data set (or borehole log) is less desirable for use in a liquefaction evaluation. The fact that the DGWT was not recorded or appears blank in the borehole log does not necessarily mean that there is no ground water present. The case may be that the borehole was too shallow to encounter ground water or the borehole caved in before a measurement was made or taking the measurement was neglected. Any of these cases associated with no DGWT recorded are poor practice, but unless there is a valid entry in the borehole log it is very difficult to make an assessment of the actual DGWT. On the other hand, if there is a value for the DGWT in the boring log, it is known that an observation well (or piezometer) is more appropriate than making a DGWT measurement during the drilling process. These different cases mentioned will be associated with different levels of quality and can be accounted for quantitatively, again by the use of weighing factors. The data types available in the borehole log can be evaluated according to their significance in liquefaction assessment as presented for the DGWT above. Table 4 summarizes the significance of some common data types in a borehole log obtained for liquefaction assessment.

D'Andria, et al. (1995) introduced the concept of a quality attribute table that accounts for the relative importance of each data item. The quality attribute table was specifically designed for the borehole data quality evaluation for liquefaction analysis and a different table should be developed for each purpose. The method of assigning different weights or scores for each attribute found in a borehole log can be considered a way to quantify the relative quality of the logs depending on their use. This concept is very similar to the "data quality model" proposed by Venkataramanan (1996), where models were developed for different geotechnical analysis applications including liquefaction.

51

Table 4 - Data Types and their Significance in Liquefaction Assessment
(adapted from D'Andria, et al., 1995)

Data Type	Significance
Date	Date on which the boring was performed.
	a) Temporal nature of the data (e.g. an earthquake might have changed the subsurface conditions).
	b) Indicates state-of-practice at the time of investigation. (e.g., year, 1950's vs 1980's)
Depth of Ground Water Table (DGWT)	DGWT may have been recorded at the time of drilling, 24 hours after drilling or may be monitored by an observation well for a longer period of time. There may also be some implications associated with the depth of the borehole.
Soil Type	Uniformly graded materials are more susceptible to liquefaction than well graded materials (Seed, 1971). Fine sand tends to liquefy more easily than the coarse sand, gravely soils, silts, or clays.
Relative Density or Void Ratio	The liquefaction susceptibility is determined to a high degree by its void ratio or relative density (Seed, 1971).
N-Value	Liquefaction potential is evaluated using resistance to penetration, and a corrected N-value form a SPT is used.

Application to the Treasure Island Site

The quality ranking system proposed by D'Andria et al., (1995) above was utilized in a collection of about 300 borehole logs in a pilot study site located in Treasure Island, CA (Luna, 1995). This site is a Naval Station in the San Francisco Bay Area which is in the process of being decommissioned. This man-made island was built in the late 1930s (hydraulic fill contained in dikes with rock revetment) there was strong evidence of liquefaction during the 1989 Loma Prieta earthquake. Since 1941 the Navy has had jurisdiction of the Island and is responsible for overseeing the work of all the geotechnical consultants directing subsurface investigations at the site. Engineering reports, plans and borehole data were made available by the Western Naval Facilities Command (NAVFAC). The pilot study used data collected from this site in the last 50 years and was implemented in a spreadsheet to aid in data entry, display and manipulation to finally perform the quality ranking of borehole logs.

The borehole logs were subjected to the criteria developed in the quality attribute tables developed for liquefaction. Their corresponding scores were used to make relative comparisons between the borehole logs out of a maximum score of 71; an ordered bar chart of the scores is shown in Figure 2. It was observed that some of the oldest data scored low and the data collected by specialized consultants scored high, however, some exceptions were noted. The lowest score (19) was performed in 1967 and it had no record of depth to groundwater table, drilling date, or any information pertaining the equipment used. On the other hand, the highest score (42) was for a borehole log conducted for liquefaction evaluation purposes in 1987 which contained most of the data types shown in Table 3. This quantitative relative quality can be used to group data by similar levels of quality and perform separate analysis, such as using a subset of the data discriminating the low quality information.

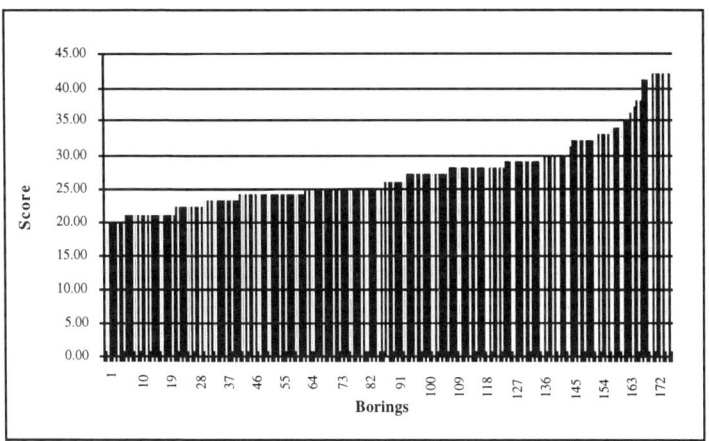

Figure 2 - Borehole Log Quality Evaluation for the Treasure Island site
(after Luna, 1995 and D'Andria, et al., 1995).

At the Treasure Island pilot study site about 40 percent of the data was ignored in the development of the database since the data ranked too low in the data quality evaluation process. The spatial liquefaction analyses were run on data considered to be of medium and high quality based on the score obtained (greater than 20). This meant reducing the database used in the GIS from an original borehole log count of 300 to 178.

Conclusion

Systematic and quantitative means to evaluate geotechnical data quality as it relates to geotechnical earthquake engineering were presented. Evaluating the quality of data used in spatial information systems provides an added value to the engineering analysis and more confidence in the results.

Researchers and practitioners have the responsibility to address the data quality issues when data will be electronically available in a spatial database. As computer technology advances, the tendency of engineering analysis based on data obtained from a spatial databank will become more prevalent. It is the opinion of the author that the use of these spatial information systems cannot progress to a sophisticated engineering level if the issues of data quality are not addressed in a systematic fashion.

Acknowledgments

Partial funding was provided by the U.S. Navy NFESC and the National Science Foundation grants (CMS-9457549 and CMS-9615707) and their financial support is gratefully acknowledged.

53

References

Aalders, J.G.L, et al. (1995), "Data Quality", Proceedings of the Specialist Meeting of GISDATA, European Science Foundation, Portugal.

Buzan, T., (1991), "Use Both Sides of Your Brain", Third Edition, Plume Publishers, 154 pp.

D'Andria, G., J.D. Frost, R. Luna, and E.J. Macari (1995), "Quality of Borings for Use in Liquefaction Spatial Analysis", Proc. of the 10th PanAmerican Conference on Soil Mechanics and Foundation Engineering, Guadalajara, Mexico, Vol 3, pp. 1739-1749.

GeoInfo Systems (1994), "NSDI Forum", Geo Info Systems, November/December, pp. 14.

Jamiolkowski, M., et. al (1988), "New Correlations of Penetration Tests for Design Practice", Penetration Testing 1988, Vol. 1, Proceedings of ISOPT-1 (Orlando), Balkema, Rotterdam, pp. 263-296.

Long, J.H. (1994), "Databases for Deep Foundations", Proceedings of the Ohio Valley Soils Seminar, October 21, 1994, Lexington, KY, pp. 18.

Luna, R. (1995), "Liquefaction Evaluation Using A Spatial Analysis System", Ph.D. Thesis submitted to the Faculty of the Georgia Institute of Technology, March, 309 pp.

Martin, J.R., and Pond, E.C. (1996), "Seismic Parameters for the Central United States Based on Paleoliquefaction Evidence in the Wabash Valley", Virginia Polytechnic Institute and State University, Geotechnical Engineering, Blacksburg, VA, Research Contract No. 14-08-001-G2141, August, 581 pp.

Ohta, Y. and Goto, N., (1978), "Empirical Shear Wave Velocity Equations in Terms of Characteristic Soil Indexes", Earthquake Engineering and Soil Dynamics, Vol. 6, pp. 167-187.

Riggs, Charles O.,(1986), "North American Standard Penetration Test Practice: An Essay", Use of In Situ Tests, pp. 949-967.

Rockaway, T, Frost, J.D., Eggert D.L., Fehlenberg, S.L., (1995) "Geotechnical Earthquake Hazard Analysis of The Evansville, Indiana Area", 5th International Conference on Seismic Zonation, Nice, October, Vol. II, pp. 1351-1358.

Rockaway, T.D., Frost, J.D., Eggert, D.L., and R. Luna, (1997) "Spatial Earthquake Hazard Assessment of Evansville, Indiana", USGS Seismic Research Letters - Illinois Basin Special Topic Report, in press.

Sarasua, W.A. and Nissalke, T., (1993), "State of the Art in Spatial Data Transfer: FIPS 173", Proceedings of the NSF Workshop on Geographic Information Systems and their Application in Geotechnical Earthquake Engineering, ASCE, pp. 95-99.

Seed, H.B., and Idriss, I.M., (1971), "Simplified Procedure for Evaluating Soil Liquefaction Potential", *Journal of Soil Mechanics and Foundations Division*, ASCE 97(SM9), pp. 1249-1273.

Skempton, A. W., 1986, "Standard Penetration Test Procedures and the Effects in Sands of Overburden Pressure, Relative Density, Particle Size, Aging, and Overconsolidation", Geotechnique, 36, No. 3, pp. 425-447.

Threadgold, L., and Hutchinson, R.J., (1992), "The Electronic Transfer of Geotechnical Data from Ground Investigations", Geotechnique Et Informatique, Proceedings of the International Conference of Geotechnics and Computers, Paris, pp. 749-756.

Troll, D.G. (1994), The Role of a Knowledge Based System in Interpreting Geotechnical Information", Geotechnique.

Venkataramanan, C, (1996), "A System for Data Quality Evaluation Applied to Borehole Logs", MS Thesis, Georgia Institute of Technology, 173 pp.

Wycoff, J., (1991), "Mindmapping - Your Personal Guide to Exploring Creativity and Problem-Solving", Berkeley Books, New York, 173 pp.

Spatial Ground-Motion Amplification Analysis

Roger D. Borcherdt[1]

Abstract

Modern Geographic Information System (GIS) technology provides a powerful tool for the compilation, archival, analysis, and display of geographic earth science information necessary for spatial analyses of ground motion amplification. Geologic, seismic, and physical property data bases, as compiled in a GIS for the San Francisco Bay region, are used to illustrate the application of GIS to the preparation of predictive GIS maps of amplification capability for strong ground shaking. These maps are consistent with new provisions recently adopted for U.S. building codes. They are used in conjunction with predictive maps of ground shaking opportunity to prepare predictive maps of ground shaking potential.

Introduction

Predictive maps of strong ground shaking are the basis for earthquake hazard mitigation. They are needed for a variety of reasons, including earthquake-resistant building-code provisions, earthquake loss estimates, emergency response plans, and public policy decisions. Modern Geographic Information Systems (GIS) afford important new opportunities to compile, archive and update predictive maps based on extensive spatial data bases and modern ground motion attenuation relationships.

The great 1906 earthquake and many subsequent earthquakes affecting areas such as Mexico City, Loma Prieta, and Kobe have demonstrated that strong ground shaking and resultant amounts of damage depend strongly on the geological character of the ground. In general, certain period bands of ground motion may be amplified significantly by various types of geologic deposits, with the amount of amplification for periods near 1 second increasing with softness of the deposits (see e.g. Lawson, 1908; Borcherdt, 1971, Seed, et al., 1992, and Borcherdt and Glassmoyer 1992, Borcherdt, 1995c). This report summarizes methodologies for spatial analysis of regional data sets to prepare predictive GIS maps of ground

----

[1]United States Geological Survey, 345 Middlefield Road, Menlo Park, CA 94025

shaking amplification which in turn are used in preparation of final predictive maps of ground shaking.

Ground shaking maps may be prepared showing geographic variations in any of several different parameters inferred from strong-motion recordings or in terms of earthquake intensity inferred from damage observations. Maps prepared in terms of spectral response for specific response periods can be useful for site specific design considerations. Maps showing earthquake intensity can be useful for various public policy decisions and some emergency response considerations. Consequently, amplification capability and predicted ground shaking maps prepared for an area need to be interpretable in terms of both earthquake intensity and ground-motion parameters in a self-consistent manner. The methodology for derivation of such a set of relations for the San Francisco Bay region is described.

Predictive GIS Mapping of Ground-Shaking Amplification
(Amplification Capability Maps)

Characterization of potential variations in strong ground shaking amplification in an urbanized region requires earth-science data available throughout the region of interest. In most regions, data most readily available is geologic data presented in the form of geologic maps. Most geologic maps, however, are compiled for purposes other than estimating ground response. For example, most geologic maps differentiate bedrock units in considerable detail, but only crudely differentiate young, unconsolidated sedimentary deposits of most concern for estimates of ground response.

Extensive programs to collect borehole, seismic velocity and geologic logs in all major sedimentary units have been conducted in the last two decades in the San Francisco Bay and Los Angeles regions. These programs have yielded well defined relations between observed amplification of shaking as a function of period, shear-wave velocity, physical properties and mapped geologic units (Borcherdt, et al., 1978, Fumal, 1978, and Fumal and Tinsley, 1985). An important correlation, which permits site specific information to be extrapolated to a regional scale is that between shear velocity and physical properties of the units, especially grain size for soils and fracture spacing for rock. This correlation is summarized by Borcherdt (1994b).

Definitions of the site classes, as proposed for the new NEHRP site classes, are given in Table 1(Borcherdt, 1994b). These definitions are based on recent comprehensive sets of in-situ data collected to determine relationships between mappable properties of near-surface materials, shear-wave velocity, and ground-motion amplification (Borcherdt et al., 1978, Fumal, 1978; Fumal and Tinsley, 1985; Borcherdt, et al., 1991). These in-situ data derived from detailed borehole logs are published for about 130 sites in the San Francisco and Los Angeles regions (Fumal, 1978; Borcherdt et al., 1978; Gibbs et al., 1975; 1976; 1977; 1980; Fumal, et al. 1981, 1982, 1984, Fumal and Tinsley, 1985; Gibbs et al., 1992, and Fumal, 1991).

57

TABLE 1 -- Site classes proposed for building-code provisions (NEHRP, UBC), amplification factors, and intensity increments.

SITE CLASS	CLASSIFICATION CRITERIA	Shear-wave Velocity			Thickness	Amplification Factors		Intensity Increments	Amplification Capability
Name	General Description	min m/s	avg m/s	max m/s	min m	F_a	F_v	δI_{MM}	
SC-I	**FIRM and HARD ROCKS**								
SC-Ia Ao*	HARD ROCKS (e.g. metamorphic rocks with very widely spaced fractures).	1400	1620			0.9	0.8	-0.4	
SC-Ib A	FIRM to HARD ROCKS (e.g. granites, igneous rocks, conglomerates, sandstones, and shales with close to widely spaced fractures).	700	1050	1400		1.0	1.0	0.0	None to low
SC-II B	GRAVELLY SOILS and SOFT to FIRM ROCKS (e.g. soft igneous sedimentary rocks, sandstones, and shales, gravels, and soils with > 20% gravel).	375	540	700	10	1.3	1.5	0.7	Low to intermediate
SC-III C	STIFF CLAYS and SANDY SOILS (e.g. loose to v. dense sands, silt loams and sandy clays, and medium stiff to hard clays and silty clays (N>5 blows/ft)).	200	290	375	5	1.6	2.3	1.3	Intermediate to high
SC-IV D	SOFT SOILS								
*SC-IVa*D1	NON SPECIAL-STUDY SOFT SOILS (e.g. loose submerged fills and very soft to soft (N<5 blows/ft) clays and silty clays < 37 m (120 ft) thick).	100	150	200	3	2.0	3.5	1.9	High to very high
SC-IVb E	SPECIAL-STUDY SOFT SOILS^^ (e.g. liquefiable soils, quick and highly sensitive peats, highly organic clays, very high plasticity clays (PI>75%), and soft soils more than 37 m (120ft) thick).				3				

(^) Mean shear velocity to a depth of 30 m (100 ft). (^^) Site-specific geotechnical investigations recommended for this class.
(*) Initial site class designation (Martin, 1994) not recommended for use in code revisions due to confusion with letter designations for seismic performance

Short- and mid-period amplification factors, as recently introduced for revised earthquake resistant building code provisions, are given as a function of the shear-wave velocity to a depth of 30 m and input ground-motion level (Borcherdt, 1994b) by:

$$F_a = (\frac{v_o}{v_s})^{m_a} \qquad \qquad (1a)$$

and

$$F_v = (\frac{v_o}{v_s})^{m_v} \qquad \qquad (1b)$$

where,

1) v_s is the mean shear-wave velocity to 30 m (100 ft) and may be either inferred from physical properties or measured directly at the site,

2) v_o is the average shear-wave velocity for the site class chosen as the reference ground condition, and

3) m_a and m_v are empirically derived exponents, implied by the amplification factor for the Soft-soil site class (SC-IV; E) specified at the 0.1g input ground-motion by strong-motion data and at higher levels by extrapolation using numerical modeling results.

Corresponding relations for Modified-Mercalli intensity increments (δI) as inferred from the 1906 earthquake in terms of the amplification factors (F_a, F_v) and the ratios of the shear wave velocities for the reference site condition and that at the specified site or site class, (Borcherdt, et al., 1995) is

$$\delta I_{MM} = 3.48 \, Log_{10}[F_v] = 3.48 Log_{10}[F_a^{\,m_v/m_a}] = 3.48 \, Log_{10}[\frac{v_o}{v_s}]^{m_v} . \qquad (2)$$

Short- and mid-period amplification factors and intensity increments for each site class can be predicted by (1) and (2) using the midpoint of the shear-wave velocity interval. The values shown in Table 1 are computed for an input ground motion level near 0.1 g with $m_a = 0.35$ and $m_v = 0.65$. Amplification capabilities consistent with these values are ascribed to each unit (Borcherdt, et al., 1991).

The site class definitions, amplification factors, and intensity increments as specified in Table 1 and by equations (1) and (2) provide the basis for preparing maps which delineate areas according to their capability to amplify input ground shaking. As an example, the digital geologic GIS data base compiled by Wentworth (1993) is used to prepare an amplification capability map for the San Francisco Bay region (Figure 1).

Physical property attributes associated with each unit (polygon) in the geologic data base were used to classify the 43 digital geologic units into the distinct NEHRP site class designations. The resultant site class map is shown in Figure 1. It delineates five NEHRP site classes in the San Francisco Bay region.

59

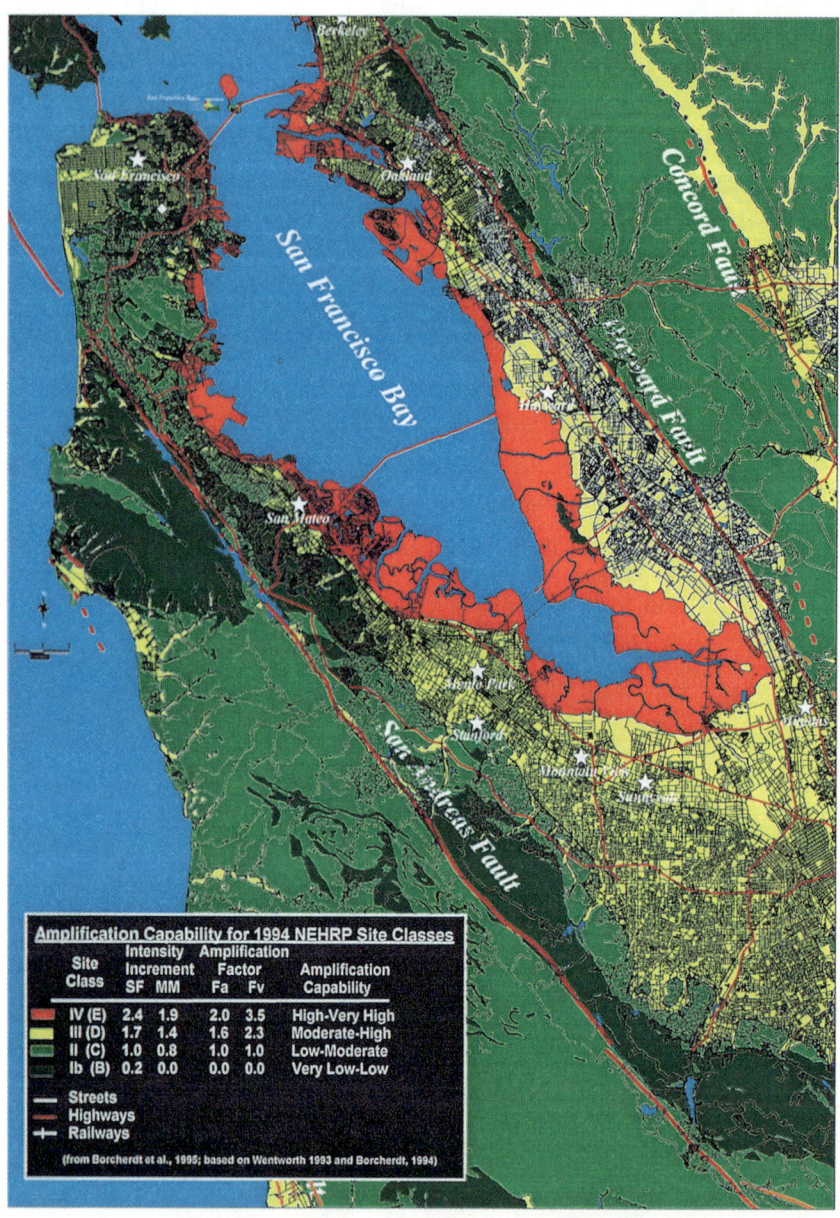

Figure 1. Map showing NEHRP site classes and corresponding amplification capability as specified by average amplification factors Fa and Fv and corresponding intensity increments specified in terms of shear-wave velocity.

The site class map also may be interpreted as a ground-motion amplification map or a map showing amplification capability with the assignment of the associated amplifications, intensity increments and amplification capabilities as specified in Table 1. However, if the map is to be used to infer actual average ground-motion amplification levels then for complete consistency with NEHRP procedures the values assigned to units near the earthquake source of interest should be adjusted for input ground motion level. This adjustment will be of most significance for the short-period amplification factor for NEHRP site class E.

The amplification capability map is derived from maps showing only the distribution of materials as mapped at the surface. Consequently, the map should only be interpreted to indicate the specified amplification levels and associated intensity increments if deposits are sufficiently thick. If the mapped deposits are in general less than 3 meters, the map will tend to over estimate the amplification. The map is of special use for identifying areas for which amplification may be a problem and may require further investigation to determine actual thickness of soil deposits.

<u>Predictive GIS Mapping of Ground Shaking for a Uniform Ground Condition</u>

The opportunity for strong ground shaking to exceed some specified level depends on characteristics of the earthquake source, crustal attenuation, local geologic conditions and occurrence frequency. Maps prepared to account for all of these influences except the local site conditions are termed sometimes ground shaking opportunity maps (Youd and Perkins, 1978; Borcherdt et al., 1991) or ground shaking maps for a uniform ground condition.

Regression relations describing the attenuation of various ground motion parameters determined from strong-motion recordings have been derived by several investigators. As an example, the empirical relation derived by Boore, Joyner, and Fumal (1993; 1994a; 1994b) predicts the peak and random components of acceleration, velocity, and spectral ordinates as a function of earthquake magnitude, rupture type, distance, and site conditions. This empirical relation is

$$Log_{10}[Y(M_w, v_S)] = C(M_w, v_S) + B_{SA} + b_4 r + b_5 Log_{10}[r]$$ (3a)

where

$$C(M_w, v_s) \equiv b_S + b_2(M_w - 6) + b_3(M_w - 6)^2 - b_V Log_{10}[\frac{v_A}{v_s}]$$ (3b)

$Y \equiv$ a random or peak component of ground acceleration in units of g or an ordinate of spectral acceleration in units of cm/sec,

$$b_S \equiv \begin{cases} b_{SS} & \textit{for a Strike–Slip Source} \\ b_{RS} & \textit{for a Reverse–Slip Source} \end{cases}$$

or

61

$b_S \equiv b_{1all}$ from the combined data set for all sources   (3c)

$M_w \equiv$ Moment Magnitude,   (3d)

$$r \equiv \sqrt{d^2 + h^2} ,$$ (3e)

$d \equiv$ distance from surface projection of crustal rupture zone in kms,   (3f)

$v_s \equiv$ shear-wave velocity to a depth of 30 m at the site,   (3g)

$B_{SA} \equiv$ response spectral acceleration - response spectral velocity, and   (3h)

$b_S, b_2, b_3, b_4, b_5, b_V,$ and V_A are regression coefficients.   (3i)

Numerical values for the regression coefficients (1i) for the peak and random components of ground acceleration and the spectral ordinates at 0.3 and 1.0 second are summarized in Table 2.

Table 2 -- Summary of regression coefficients inferred for horizontal components of peak acceleration and spectral ordinates at periods of 0.3 and 1.0 seconds (from Boore, Joyner, and Fumal, 1993, 1994a, and 1994b).

	b_{ss}	b_{RS}	b_{1ALL}	b_2	b_3	b_4	b_5	B_{SA}	b_V	h	v_A	SIG	1SIGE	SIG
Accel.														
Larger	-0.068	0.017	-0.038	0.216	0.000	0	-0.777		-0.364	5.480	1390	0.194	0.051	0.000
Random	-0.136	-0.051	-0.105	0.229	0.000	0	-0.778		-0.371	5.570	1400	0.187	0.080	0.098
PSRV														
Larger														
0.3 sec	2.063			0.354	-0.920	0	-0.902	-1.670	-0.372	5.790	2340			
1.0 sec	1.858			0.444	-0.016	0	-0.825	-2.193	-0.641	2.870	1530			
Random														
0.3 sec	1.930	2.019	1.974	0.334	-0.070	0	-0.893	-1.670	-0.401	5.940	2130	0.191	0.021	0.120
1.0 sec	1.701	1.755	1.724	0.450	-0.014	0	-0.798	-2.193	-0.698	2.900	1410	0.206	0.093	0.141

Corresponding attenuation relations have been derived for attenuation of 1906 earthquake intensity on the basis of only the most reliable earthquake intensity data. (Borcherdt, et al., 1975). These equations have been used to derive regression coefficients relating intensity to ground shaking parameters for a repeat of the 1906 earthquake (Borcherdt, et al., 1995). The resultant equation for intensity (Modified-Mercalli scale), expressed as a function of magnitude and site conditions, is,

$$I_{MM}(M_w, v_s) = \alpha_Y + \beta_Y + \gamma_Y \{C(M_w, v_s) + B_{SA} + b_4 r + b_5 Log_{10}[r]\} . \quad (4a)$$

where the regression coefficients β_Y an γ_Y are given in Table 3 and the residual coefficient is defined by

$$\alpha_Y \equiv I_{MM}(M_w, v_s) - \{\beta_Y + \gamma_Y\ Log_{10}[Y(M_w, v_s)]\}, . \qquad (4b)$$

and for the 1906 earthquake with $M_w = 7.7$ and $v_s = 1050$ m/s for firm to hard rock (site class Ib(B)), is from Borcherdt, et al., (1995), given for peak acceleration by

$$\alpha_{pga} = -0.556 - 1.495\ Log_{10}[d] + 1.816\ Log10\sqrt{d^2 + 5.48^2}, \qquad (4c)$$

for the random component of the response spectral acceleration at 0.3 sec by

$$\alpha_{rsa\ rc@\ 0.3} = -4.075 - 1.494\ Log_{10}[d] + 1.849\ Log10\sqrt{d^2 + 5.94^2}, \qquad (4d)$$

and for the random component of the response spectral acceleration at 1.0 sec by

$$\alpha_{rsa\ rc@\ 1.0} = -4.701 - 1.495\ Log_{10}[d] + 1.628\ Log10\sqrt{d^2 + 2.9^2}. \qquad (4e)$$

Table 3 Regression coefficients (β and γ) for Modified-Mercalli intensity expressed as a linear function of the logarithm of peak acceleration (pga) and response spectral acceleration (rsa) at 0.3 and 1.0 secs.

	pga	rsa @	
		0.3 secs	1.0 secs
β	8.3	7.6	8.1
γ	2.3	2.1	2
R^2	0.98	0.98	0.99

Equation (4) implies earthquake intensity (Modified-Mercalli scale) can be predicted exactly as a function of magnitude (M_w) and site conditions (v_s), using the empirical attenuation laws for peak acceleration or the random components of response spectral acceleration at 0.3 and 1.0 seconds derived from strong-motion measurements.

Equations 1- 4 illustrate the desired relations to prepare a self-consistent set of maps for a particular region in terms of either one of the modern ground shaking parameters such spectral acceleration at 1 second or earthquake intensity. The relations can be used to depict geographical variations in strong ground shaking as inferred from both strong motion parameters and damage observations.

Predictive GIS Mapping of Ground Shaking for Non-uniform Ground Conditions (Ground Shaking Potential Maps)

Final predictive ground shaking maps must be comprised of ground motion predictions for a uniform ground condition such as "firm to hard rock" plus predictions of the amplification effects of any near-surface deposits. The potential for strong ground shaking to exceed a specified level depends on the opportunity for the motions to exceed a particular level on a uniform ground condition and the amplification characteristics of the local geologic deposits. Consequently, maps showing ground-shaking potential represent the composite or superposition of maps showing ground shaking opportunity and amplification capability. Such maps delineate geographic variations in strong ground shaking induced by a set of non-uniform geologic ground conditions.

The empirical attenuation relations (3) and (4) when combined with equation (1) for the short- and mid-period amplification factors (F_a, F_v) and equation (2) for intensity increments provide a rigorous basis for predictive ground shaking maps. A consistent set of estimates may be derived, either in terms of measured ground shaking parameters or in terms of its effects as specified using earthquake intensities. Examples of GIS outputs for the San Francisco Bay Region based on these equations for a repeat of the 1906 earthquake are shown in Figures 2 and 3. Predictive maps for peak ground acceleration, response spectral acceleration at 1.0 secs and earthquake intensity (Modified -Mercalli scale) are shown.

Conclusions

Geographic information system technology provides an evolutionary new tool for the archival, management, and communication of large amounts of spatially linked information necessary for spatial analysis of ground shaking amplification. Predictive GIS maps derived herein provide geographic delineation of site classes consistent with new provisions being considered for U. S building codes. Predictive maps of ground shaking potential are derived in terms of spectral ordinates from modern regression relations and earthquake intensity as inferred from historic observations of damage in the region. Empirical equations, which express intensity as a function of the various ground motion parameters, allow the maps to be used for both technical design purposes and public policy decisions.

Acknowledgements

The kind invitation of the program committee to present this manuscript is appreciated. Research results summarized here have been developed as part of several cooperative scientific efforts referenced herein. The contributions of all of my previous co-authors are sincerely appreciated.

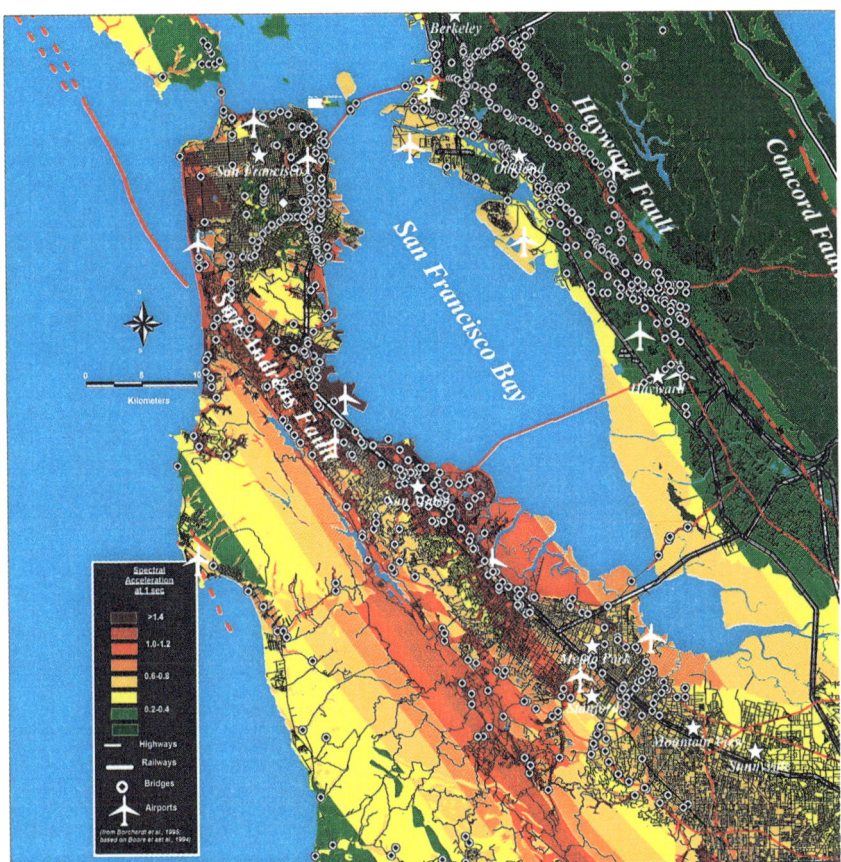

Figure 2. Map showing ground-shaking potential for a repeat of the 1906 earthquak
   (Mw = 7.7) , derived from GIS superposition of maps showing ground shaking
   opportunity and amplification capability (see Figure 1).

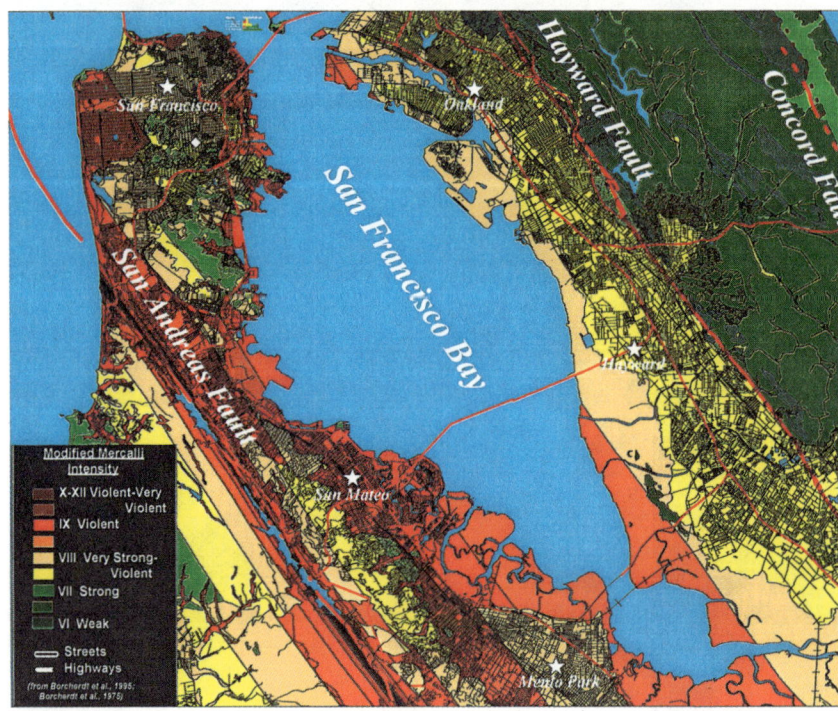

Figure 3. Map showing Modified-Mercalli intensity for a repeat of the 1906 earthquake (Mw = 7.7).

References Cited

Boore, D.M., Joyner, W.B., and Fumal, T.E., 1993, Estimation of response spectra and peak accelerations from western North American earthquakes: an interim report, U. S. Geological Open-file Report, 93-509, 72 pp.

Boore, D.M., Joyner, W.B., and Fumal, T.E., 1994a, Estimation of response spectra and peak accelerations from western North American earthquakes: an interim report part 2, U. S. Geological Open-file Report, 94-127, 40 pp.

Boore, D.M., Joyner, W.B., and Fumal, T.E., 1994b, Estimation of response spectra and peak accelerations from western North American earthquakes: an interim report part 2 with insert, U. S. Geological Open-file Report, 94-127, 4 pp.

Borcherdt, R.D., ed., 1975a, Studies for seismic zonation of the San Francisco Bay region: U.S. Geological Survey Professional Paper 941-A, 102 p.

Borcherdt, R.D., 1975b, Studies for seismic zonation of the San Francisco Bay region; a brief summary: Proc. U.S. National Conference on Earthquake Engineering, Ann Arbor, Mich., p. 123-127.

Borcherdt, R.D., 1994a, Simplified site classes and empirical amplification factors for site-dependent code provisions, *Proceedings NCEER, SEAOC, BSSC workshop on site response during earthquakes and seismic code provisions*, University of Southern California, Los Angeles, California, November 18 - 20, 1992.

Borcherdt, R.D., 1994b, Estimates of site-dependent response spectra for design (Methodology and Justification), *Earthquake Spectra*, **10**, pp 617-673.

Borcherdt, R.D., Lawson, S., Pessina, V., Bouabid, J., and Shah, H.C., 1995, Applications of geographic information system technology (GIS) to seismic zonation and earthquake loss estimation, *State-of-the-Art Lecture, Fifth International Conference on Seismic Zonation, Procs.*, Nice, France, v. **III**, *p. 1933-1973.*

Borcherdt, R.D., 1995c, Strong ground motion generated by the Hanshin-Awaji (Hyogo-ken Nanbu) earthquake of January 17. 1995 in "*The Hyogo-ken Nanbu earthquake of January 17. 1995*" J.P Moehle, ed., Earthquake Engineering Research Center Report No. UCB/EERC-95/10, University of California, Berkeley, CA, p. 11-44.

Borcherdt, R.D., and Gibbs, J.F., 1975, Prediction of maximum earthquake intensities for the San Francisco Bay region: U.S. Geological Survey Open-File Report 75-180, 41 p.

Borcherdt, R.D., Gibbs, J.F., and Fumal, T.E. 1978. Progress on ground motion predictions for the San Francisco Bay region, California, in U.S. Geological Survey Circular 807. Also, Proceedings Second International Conference on Microzonation Safer Constr. Research, Appl. 2, 1, p.13-25.

Borcherdt, R.D., and Glassmoyer, G., 1992, On the characteristics of local geology and their influence on ground motions generated by the Loma Prieta earthquake in the San Francisco Bay region, California: *Bulletin of the Seismological Society of America*, 82, p. 603-641.

Borcherdt, R.D., Wentworth, C.M., Glassmoyer, G., Fumal, T., Mork, P., and Gibbs, J., 1991, On the observation and predictive GIS mapping of ground response in the San Francisco Bay region, California: *Fourth International Conference on Seismic Zonation, Stanford, California, Proceedings*, **III**, p. 545-552.

Borcherdt, R.D., Wentworth, C.M., Glassmoyer, G., Fumal, T., Mork, P., and Gibbs, J., 1991, On the observation and predictive GIS mapping of ground response in the San Francisco Bay region, California: *Fourth International Conference on Seismic Zonation, Stanford, California, Proc., III*, p. 545-552.

California Code of Regulations, Title 14, Chapter 8, Article 10. Seismic Hazards Mapping.

Fumal, T.E. 1978. Correlations between seismic wave velocities and physical properties of geologic materials in the San Francisco Bay region, California. *U.S. Geological Survey Open-File Report 78-1067*.

Fumal, T.E. 1991. Shear-wave velocity estimates and site geology for strong-motion recordings sites of the Loma Prieta earthquake of October 17, 1989. *U.S. Geological Survey Open-File Report*, 91-311, 163 pp.

Fumal, .E., Gibbs, J.F., and Roth, E.F., 1981, In-situ measurements of seismic velocity at 19 locations in the Los Angeles, California, region: *U. S. Geological Survey Open-File report 81-399*, 123 pp.

Fumal, T.E., Gibbs, .F., and Roth, E.F., 1982, In-situ measurements of seismic velocity at 22 locations in the Los Angeles, California, region: *U. S. Geological Survey Open-File report 82-833*, 140 pp.

Fumal, T.E., Gibbs, .F., and Roth, E.F., 1984, In-situ measurements of seismic velocity at 16 locations in the Los Angeles, California, region: *U. S. Geological Survey Open-File report 84-681*, 109 pp.

Fumal, T.E. and Tinsley, J.C. 1985. Mapping Shear-wave velocities of near-surface geologic materials. *U.S. Geological Survey Professional Paper 1360*, p. 127-150.

Gibbs, J.F., Fumal, T.E., and Borcherdt, R.D. 1975. *In-situ* measurements of seismic velocities at twelve locations in the San Francisco Bay region. *U.S. Geological Survey Open-File Report 75-564*.

Gibbs, J.F., Fumal, T.E., and Borcherdt, R.D. 1976. *In-situ* measurements of seismic velocities in the San Francisco Bay region—Part II. *U.S. Geological Survey Open-File Report 75-731*.

Gibbs, J.F., Fumal, T.E., Borcherdt, R.D., and Roth, E.F., 1977. *In-situ* measurements of seismic velocities in the San Francisco Bay region-Part III. *U.S. Geological Survey Open-File Report 77-850*.

Gibbs, J.F., Fumal, T.E., and Roth, E.F., 1980, In-situ measurements of seismic velocity at 27 locations in the Los Angeles, California, region: U. S. Geological Survey Open-File report 80-399,378, 169 pp

Helley, E.J., and Lajoie, K.R., 1979, Flatland deposits of the San Francisco Bay region, California, *U.S. Geological Survey Professional Paper 943*, 88 pp.

King, S.A. and Kiremidjian, A.S., 1994, Regional seismic hazard and risk analysis through geographic information systems, *The John A. Blume Earthquake Engineering Center, Report 111*, Stanford University, 168 pp.

Lawson, A.C.,and others, 1908, The California earthquake of April 18, 1906 -- Report of the State Earthquake Investigation Commission; *Carnegie Inst. Washington Pub. 87*, 2 vols.

NEHRP Recommended Provisions for the Development of Seismic Regulations for New Buildings, *1991 edition*, Prepared by Building Seismic Safety Council for Federal Emergency Management Agency, Washington D.C., vol. I, 199 pp.

Seed, R.B., Dickenson, S.E., Rau, G.A., White, R.K., and Mok, C.M., 1992, Observations regarding seismic response analyses for soft and deep clay sites, *preprint*.

Wentworth, C.M., 1993, General distribution of geologic materials in the southern San Francisco Bay region, California, U. S. Geological Survey Digital Map Data Base, INTERNET http://wrgis.wr.usgs.gov/docs/geologic/ca/cal.

Wentworth, C.M., Borcherdt, R.D., Fitzgibbon, T.T., Mork, P., and Tarr, A.C., 1991, Application of GIS technology to seismic zonation of the San Francisco Bay region, California: *Fourth International Conference on Seismic Zonation, Stanford, California, Proceedings*, **III**, p. 537-544.

Youd, T.L., 1991, Mapping of earthquake-induced liquefaction for seismic zonation, *Fourth International Conference on Seismic Zonation, Proceedings*, Stanford, California., **I**, p.111-147.

Youd, T.L., Nichols, D.R., Helley, E.J., and Lajoie, K.R., 1975, Liquefaction potential, in *Studies for seismic zonation of the San Francisco Bay region: U.S. Geological Survey Professional Paper 941-A*, Borcherdt, R.D., ed., p A68-A74.

Youd, T.L., and Perkins, D.M., 1978, Mapping liquefaction-induced ground failure potential, *Amer. Soc. Civil Engineering jour. Geotech. Eng. Div. 104*, p. 433-466.

Spatial Liquefaction Analysis

J. David Frost[1], Daniel P. Carroll[2] and Thomas D. Rockaway[2]

Abstract

Geotechnical hazard analyses play a key role in identifying and mitigating against the potential consequences of an earthquake. Geographic Information Systems (GIS) provide an environment which is ideal for conducting earthquake hazard and risk analyses where the different geotechnical and geological hazards associated with seismic events can be evaluated and the large number of factors required to describe these hazards can be taken into account. This paper reviews recent progress in liquefaction evaluation and the implementation of these developments in GIS based hazard evaluation procedures. Examples of spatial liquefaction analysis systems are described and typical results are presented.

Introduction

Liquefaction is regarded as one of the most severe seismic hazards, and often results in catastrophic damage to structures and infrastructure, and the loss of human life. The loss of soil strength and resultant vertical and/or lateral movement of soil and structures has been observed to occur during all major earthquakes (Figures 1 and 2). It is therefore important that geotechnical engineers and others have the ability to predict and evaluate areas that can potentially liquefy during an earthquake so that measures can be taken to mitigate the consequences. Once a potentially liquefiable area has been identified, the consequences of the liquefaction can be investigated. These consequences include ground failure as a result of vertical and lateral deformation.

[1]Associate Professor and [2]Graduate Research Assistant, School of Civil and Environmental Engineering, The Georgia Institute of Technology, Atlanta, GA 30332.

Figure 2 Collapse of Roadway Caused by Liquefaction and Lateral Spreading

A Geographic Information System (GIS) provides a platform where liquefaction potential and its consequences can be spatially evaluated and geographically referenced. The unique ability of a GIS to store and process large amounts of spatially referenced data make it a valuable tool in the evaluation of earthquake hazards such as liquefaction. This paper provides an overview of liquefaction analysis procedures and their suitability for use within a GIS. The benefits of spatial liquefaction analysis are illustrated with selected examples.

Liquefaction Analysis Procedures

There is no direct method for evaluating the liquefaction potential of an in situ soil deposit. However, there are several methods which directly, or indirectly measure the parameters which control liquefaction. These methods range from relatively simple qualitative assessments to detailed quantitative calculations of the liquefaction potential.

Among the more qualitative methods are those which use mapping techniques to define liquefaction susceptibility zones (eg. Youd and Perkins, 1978). Maps containing information such as geological descriptions and seismic history are compiled to form a single map of ground failure potential zones. However, further site specific studies are required for the characterization and delineation of these zones. More recent work in this area includes the mapping of a parameter termed liquefaction severity index (LSI), which is a measure of ground failure displacements during previous earthquakes (Youd and Perkins 1987), and the mapping of liquefiable layer thickness (O'Rourke and Pease, 1997).

The quantitative methods available make use of either in-situ or laboratory measurements. Information from field observations of liquefaction during previous earthquakes is interpreted in conjunction with parameters obtained from in-situ or laboratory tests to determine a factor of safety against liquefaction at given depths within a soil profile. More focus has been placed on in-situ based methods as a result of the difficulty of obtaining undisturbed samples of loose sands for laboratory testing.

The most widely used in-situ based method is the correlation of liquefaction resistance with the standard penetration (SPT) N-value as originally proposed by Seed and Idriss (1971). This approach uses an empirical correlation to compare the cyclic stress ratio required to induce liquefaction to the cyclic stress ratio generated by a specific earthquake. A factor of safety against liquefaction is then calculated at elevations within the soil stratigraphy corresponding to the SPT test elevations. Other researchers have refined the SPT based liquefaction evaluation procedure to account for factors such as material composition, percent fines and earthquake magnitude among others (eg. Tokimatsu and Yoshimi, 1983; Seed et al., 1985; Ho and Kavazanjian, 1990; Singh, 1994; Vaid and Thomas, 1994; Erten and Maher, 1995; Evans and Zhou, 1995).

The cone penetration test (CPT) is also used in a similar manner for evaluating liquefaction resistance. Initial efforts focused on estimating an equivalent SPT N-value from correlations with cone penetration resistance, q_c, and D_{50} (eg. Robertson et al., 1983; Robertson and Campanella, 1985) so that the database of experience accumulated with SPT N-values could be indirectly accessed. Compared to cases where SPT data is available, there are still relatively few cases that provide a direct correlation between CPT values and evidence of soil liquefaction although research over the past decade has helped diminish this lack of data and led to refinements in CPT based liquefaction evaluation procedures including reduced dependence on SPT-CPT correlations (eg. Seed and deAlba, 1986; Muraleetharan et al., 1991; Olsen and Koester, 1995; Suzuki et al., 1995). One advantage of the CPT based methods are that they provide a near continuous profile with depth of the factor of safety with respect to liquefaction as opposed to the sparsely distributed values obtained with the SPT based methods. This may be particularly significant in evaluating the behavior of layered soil deposits.

The concept of a liquefaction potential index was introduced by Iwasaki et al., (1978) to provide a way to help quantify how liquefaction will affect structures on the surface as a function of the depth to, and extent of, the liquefied zones within the soil profile. For example, a 2 m layer liquefying immediately beneath a shallow footing is likely to have a more detrimental effect on the performance of the footing than if the layer was located 15 m below the footing base. The LPI provides a convenient way to convert the liquefaction potential profile with depth obtained using the traditional SPT or CPT based method to a single value at the surface by using a weighting function to assign greater importance to near surface liquefaction. This aspect will be discussed later in the context of implementing liquefaction evaluation procedures within a GIS.

A range of other liquefaction evaluation procedures based on field tests have been proposed. However, to date, they have not been fully developed or have encountered some limitations. These include the use of the dilatometer test (Reyna and Chameau, 1991; Glaser and Chung, 1995), and the vibratory cone penetrometer test (Mayne, 1997) to name a few. These methods show some promise, but limited data and understanding restrict their current use and thus they will not be discussed further herein.

Uses of Liquefaction Analysis

A prominent use of the results of liquefaction analysis studies is the creation of liquefaction hazard maps. The methods used for creating these maps range from the very general and simple to the very detailed and specific. Small scale liquefaction potential maps are produced by superimposing liquefaction susceptibility and liquefaction opportunity maps (Youd and Perkins, 1978). The criteria used to compile susceptibility maps include historic occurrences and geological mapping. Criteria required to generate an opportunity map are the location and frequency of earthquake occurrence, and magnitude-distance relations. These liquefaction potential maps are used in a variety of applications to aid in earthquake hazard mitigation, including land use planning, building codes, and financial and insurance risk evaluation.

Most of the maps produced using the techniques mentioned above present only qualitative information. The geotechnical engineer is more often concerned with a numerical assessment. The "simplified procedure" for evaluating liquefaction potential (Seed and Idriss, 1971) provides a method to create quantitative and site specific liquefaction hazard maps. Standard penetration (SPT) boring information is used in conjunction with earthquake data to determine a soil layer's factor of safety against liquefaction for a specific earthquake scenario. The Liquefaction Potential Index (LPI) method (Iwasaki et al., 1978) is a measure of the degree of severity liquefaction will have at the ground surface. The method uses the factor of safety profile with depth and a weighting function to determine the probability that liquefaction will affect structures on the ground surface. The LPI is calculated at several boring locations and geostatistical procedures can be used to create a site specific liquefaction severity contour map.

However, for engineering purposes, it is often not the actual threat of liquefaction that is of prime importance, but the damaging effects that liquefaction has on structures. Liquefaction causes differential ground displacements which are detrimental to structures located on the surface. Several methods have been developed to evaluate the amount of ground displacement which will occur as a result of liquefaction during an earthquake. Youd and Perkins (1987) introduced the concept of mapping liquefaction severity index (LSI). The LSI is defined as the general maximum ground displacement of lateral spread on liquefiable deposits and is calculated by dividing the maximum horizontal displacement, measured in millimeters, by 25. Previous earthquake data are collected and analyzed to generate probabilistic maps of the expected LSI.

Two other more rigorous methods provide a means for calculating the expected ground surface vertical and horizontal displacement at specific boring locations. Both methods rely on the Seed and Idriss "simplified procedure" to first

74

evaluate the factor of safety against liquefaction for a soil layer. Once it is determined that a soil layer is subject to liquefaction (FS ≤ 1), the vertical and horizontal displacements for that layer can be calculated. For example, Castro (1987) has developed an empirical method for estimating the earthquake induced vertical settlement. The method estimates the volumetric compression of each soil layer based on correlations between cyclic shear strain and volumetric strain. The volumetric compression is obtained by multiplying the volumetric strains by the thickness of the soil layer. The total vertical settlement at the surface is the sum of the contributions of each layer. If a statistically significant number of borings are present, geostatistical methods can be used to generate contour plots of vertical deformation for the specific study area.

Empirical approaches can also be used to estimate the amount of horizontal displacement, or lateral spread, that will occur during earthquake induced liquefaction. Bartlett and Youd (1992) collected lateral spread case history data and performed regression analyses to develop empirical models to estimate the quantity of lateral spread that is likely to occur within a liquefiable layer. Two separate equations are required; one for a free-face model for areas near a steep bank, and a ground slope model for areas with gently sloping terrain (slope ≤ 6%). The models incorporate statistically significant variables such as earthquake magnitude, horizontal distance from seismic energy source, average mean grain size, average fines content, and layer thickness. The total lateral spread at the surface is obtained by summing the contributions of each soil layer.

The Role of a GIS in Liquefaction Analysis

There are many benefits to working in the spatial environment of a geographic information system where data can be geographically referenced. A GIS has the ability to store large amounts of spatial data in a much more compact and efficient form than paper maps, forms, and reports. Within a GIS, this data can also be easily retrieved through queries and the database is readily expanded and updated. Inferential techniques are also an integrated part of the GIS. Geostatistical methods such as linear interpolation, inverse distance weighing, and kriging can be used to infer values at locations where no data is available. This tool is particularly appealing to geotechnical engineers because it eliminates the need to assume an average soil profile. Results can be obtained rapidly at several locations and values can be inferred at points in between. Furthermore, engineering procedures can be embedded and integrated within the GIS in order to expand the systems capabilities. These procedures can generate new levels of information that model physical phenomena in the same spatial environment in which they occur. (Luna and Frost, 1995).

A powerful aspect of a GIS is the ability of the system to display results geographically. For example, Figure 3 shows the contours of liquefaction potential from a typical spatial analysis study. Information gained through engineering computations can immediately be viewed within its spatial context. Data such as liquefaction potential or earthquake induced settlement can be spatially overlain with maps showing locations of population density and/or critical structures.

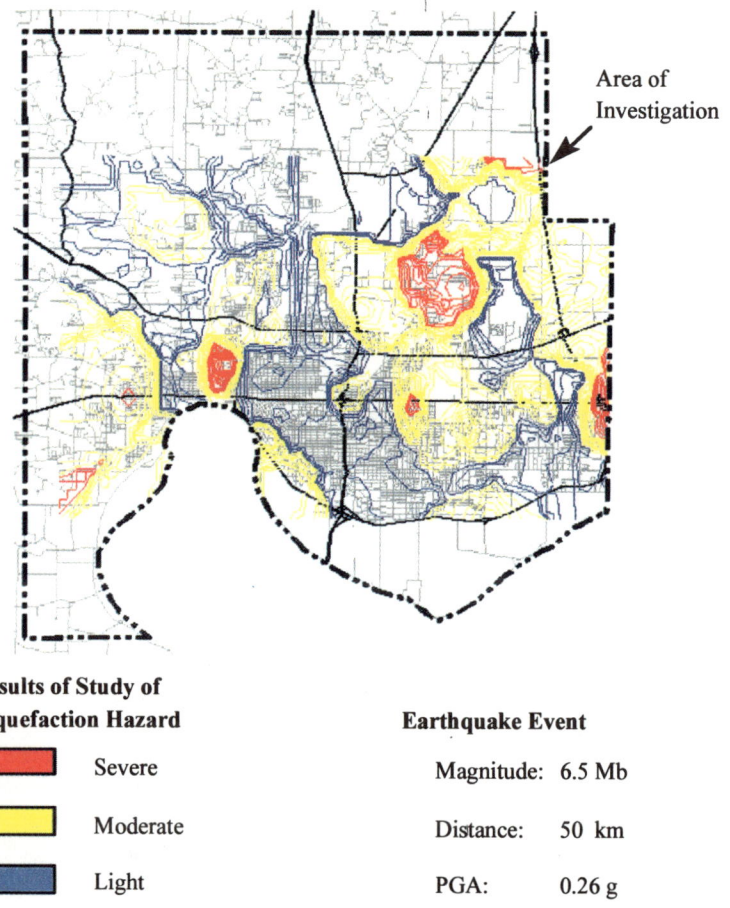

**Results of Study of
Liquefaction Hazard**

🟥	Severe
🟨	Moderate
🟦	Light

Earthquake Event

Magnitude:	6.5 Mb
Distance:	50 km
PGA:	0.26 g

Figure 3   Contours of Liquefaction Hazard

A GIS coverage can be created (Figure 4) to "hot-link" locations of photographs of structures with their geographic location. Subsequent overlaying of this photograph coverage with liquefaction contours such as those shown in Figure 3 can aid in the evaluation of the consequences of seismic activity. The spatial analysis routines within a GIS allow for numerous possibilities to compare, contrast, and evaluate results.

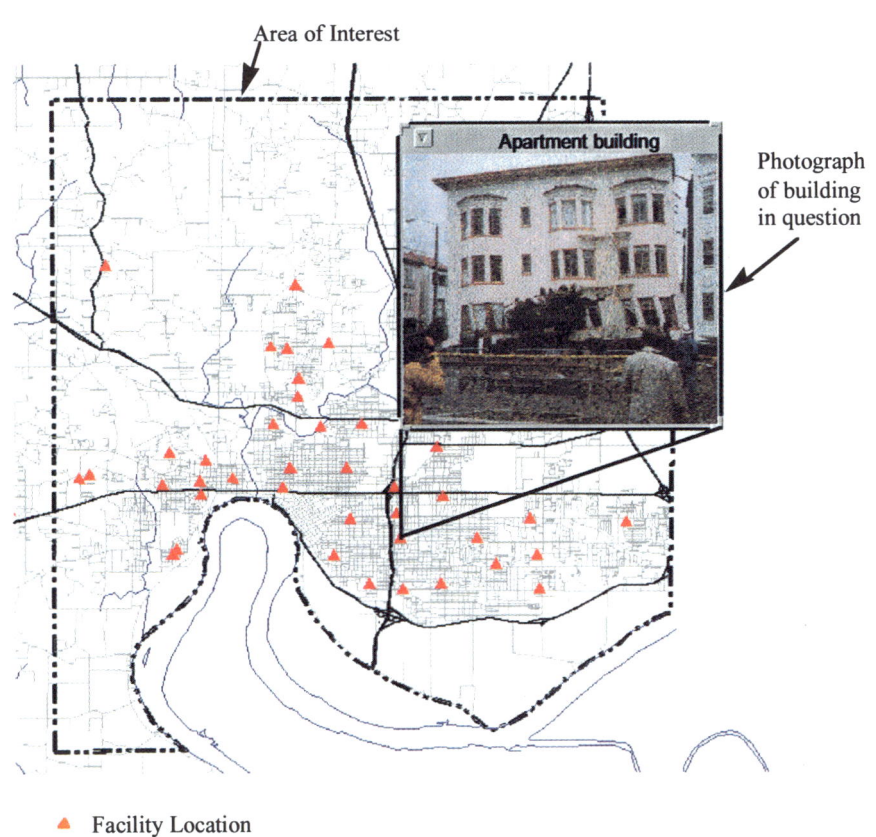

Area of Interest

Apartment building

Photograph of building in question

▲ Facility Location

Figure 4 GIS Coverage Linking Photographs to Study Area Locations

Certain methods of liquefaction analysis are more suitable for use within a GIS. For instance, when performing a regional assessment at a small scale (1:250,000) it is best to generate liquefaction potential maps by compiling susceptibility and opportunity maps from historic, geologic, and seismic data. These data sets are typically available only at small scales and show the general trends over large areas.

However, the geotechnical engineer is more often interested in obtaining quantitative results at a specific site, and thus methods which lead to the generation of detailed site specific hazard maps are more desirable. Since the majority of information obtained during a site geotechnical investigation is in the form of borehole soundings (SPT and CPT), it is ideal to use a method of liquefaction analysis which bases its calculations on data retrieved from these soundings. Seed and Idriss's "simplified procedure" is the most widely used liquefaction analysis method, and it makes direct use of the engineering parameters obtained from borehole soundings. As mentioned previously, the method determines the factor of safety against liquefaction for each soil layer at a given borehole location. Unfortunately, a GIS only has the ability to deal with two dimensional data, whereas the results from the "simplified procedure" are in a three dimensional format (factor of safety with depth profiles at specific locations on the surface). This difficulty can be overcome by incorporating the Liquefaction Potential Index suggested by Iwasaki et al. into the analyses. The LPI converts the factor of safety profiles into single point data which is ideally suited for use within a GIS framework. The use of Seed and Idriss's "simplified procedure" along with the addition of the LPI is a most effective means of performing a liquefaction hazard assessment within a GIS.

The methods previously mentioned for estimating earthquake induced ground displacements are also ideally suited for use within a GIS. These methods internally convert the displacement profile with depth to a single value at the ground surface which can be readily used in spatial analysis. It should be noted that the only limitation involved with using a GIS lies in the conversion of the 3D data to 2D using the LPI. However, in most cases, the 3D liquefaction data is used to calculate the earthquake induced settlements directly. The 3D to 2D conversion must only be performed if liquefaction susceptibility maps are to be produced. If the real interest is in the ground displacements, which is true in most cases, then the need for the conversion using the LPI is eliminated.

Example of GIS based Liquefaction Analysis

Spatial or GIS based liquefaction hazard analysis systems have been developed by a number of organizations and researchers. In some cases, liquefaction is only one of a number of hazards evaluated. The scales at which the various systems are designed to be used varies, and this dictates many of the characteristics of the database requirements and the analysis methodologies used.

For example, the National Institute of Building Sciences (NIBS), with funding from the Federal Emergency Managament Agency (FEMA), has developed a system called HAZUS, which is a GIS based system for earthquake loss estimation. The system is design to perform hazard analyses at a regional scale and thus relies of geological maps and other regionally defined data sources as opposed to site specific soundings. HAZUS is intended to be a corner-stone of FEMA's programs to promote earthquake hazard mitigation and to help local governments prepare for, and reduce the impact of earthquakes. The FEMA HAZUS loss estimation methodology is a software program that uses analytical routines and databases of information about building inventory, local geology and the location and size of potential earthquakes, as well as economic data, and other information to estimate the losses from potential earthquakes. Using a hypothetical earthquake, HAZUS estimates the severity of ground shaking, the number of buildings damaged, the number of casualties, the amount of damage to transportation systems, the extent of disruptions to electrical and water utilities, the number of people displaced from their homes, and the estimated cost of repairing projected damage and other effects.

Similarly, researchers at Stanford University have developed a GIS-based damage and loss estimation methodology which has been used to conduct earthquake hazard studies including ground motion, liquefaction, bridge classification and building damage assessments among others for a range of earthquake scenarios (eg. King et al., 1995).

A spatial earthquake hazard assessment system (GIS-QUAKE) has been developed by researchers at the Georgia Institute of Technology (Luna and Frost, 1995; Rockaway and Frost, 1997). The system expands a standard GIS into a spatial geotechnical engineering analysis tool by integrating a functional database with the spatial components of a GIS and independent engineering hazard analysis routines for assessing liquefaction, ground motion amplification and earthquake induced landslide potential. The hazard analyses are performed on point specific data and geostatistical methods are used to interpolate the results across the study area. The GIS-QUAKE architecture separates the action of the system into major components and modules. The major components identify the major tasks performed by the system while the modules define specific techniques or data critical to the component (Figure 5). In this manner, the user is provided with a variety of data and analysis techniques which enable customized analytical studies. As an example, the spatial analysis component consists of a linear interpolation module and a kriging module. The user may select either of the techniques to interpolate the results.

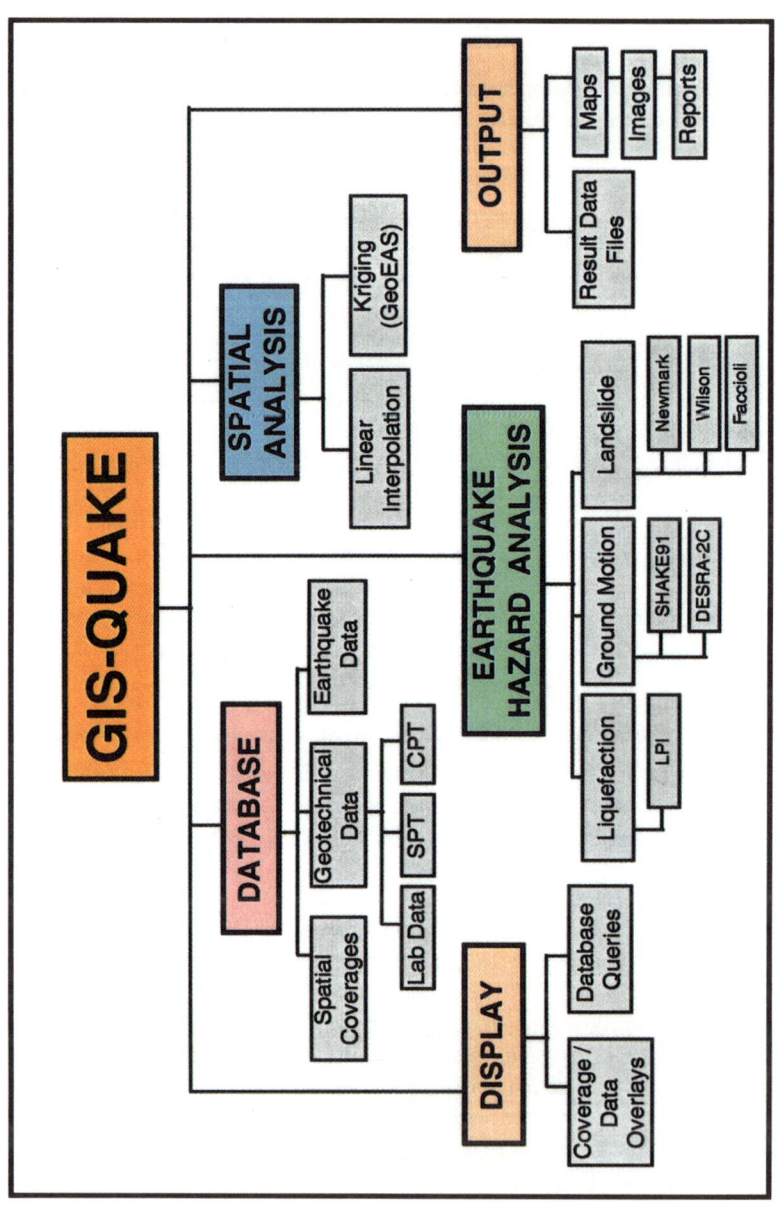

Figure 5 Architecture of GIS-QUAKE Hazard Analysis System

The results obtained with the GIS-QUAKE system are dependent on decisions of the user performing the analysis and on the amount and quality of information in the database. The system has been designed to be used by an individual who has a working knowledge of the parameters and importance of the variables used in the individual analyses. The user is expected to take an active role in the analyses and the interpretation of the results. Secondly, in order to generate high quality results, the database must be of adequate size and nature to support the specific hazard analysis for the study area. An incomplete or inadequate database can lead to gross inaccuracies in the hazard results and thus it is the duty of the operator to recognize the deficiencies and limitations of the results.

While full-functional hazard analysis systems such as those described above permit spatial liquefaction evaluation among other hazards, more specific liquefaction hazard analysis systems can be readily developed. A system is currently under-development for the U.S. Navy which combines a GIS with analysis routines for performing liquefaction, vertical settlement (Castro, 1987) and lateral deformation (Bartlett and Youd, 1992) as discussed earlier. The algorithmic procedure for performing this integrated hazard mapping is shown in Figure 6.

Conclusions

The use of spatial or GIS based systems for performing liquefaction evaluations has increased significantly over the past 5 years. The ability to combine data of different types, both qualitative and quantitative, as a result of their geographic or spatial relationships has opened new opportunities for interpreting the consequences of seismic activity. For example, the combined images in Figure 7 show how point-specific coverages of data such as geotechnical borings can be combined with aerial photographic images and site specific photographs of earthquake induced liquefaction damage to facilitate simultaneous examination and evaluation of information.

On the other hand, there is clearly a need for continued research and development. Issues such as spatial data quality, geologically sensitive interpolation techniques and required data density must be investigated. In addition, despite the apparent simplicity of many of these spatial analysis systems, there is clearly a need to ensure that those performing the analyses have appropriate training in the technical disciplines which are an integral component of the analysis techniques used in system.

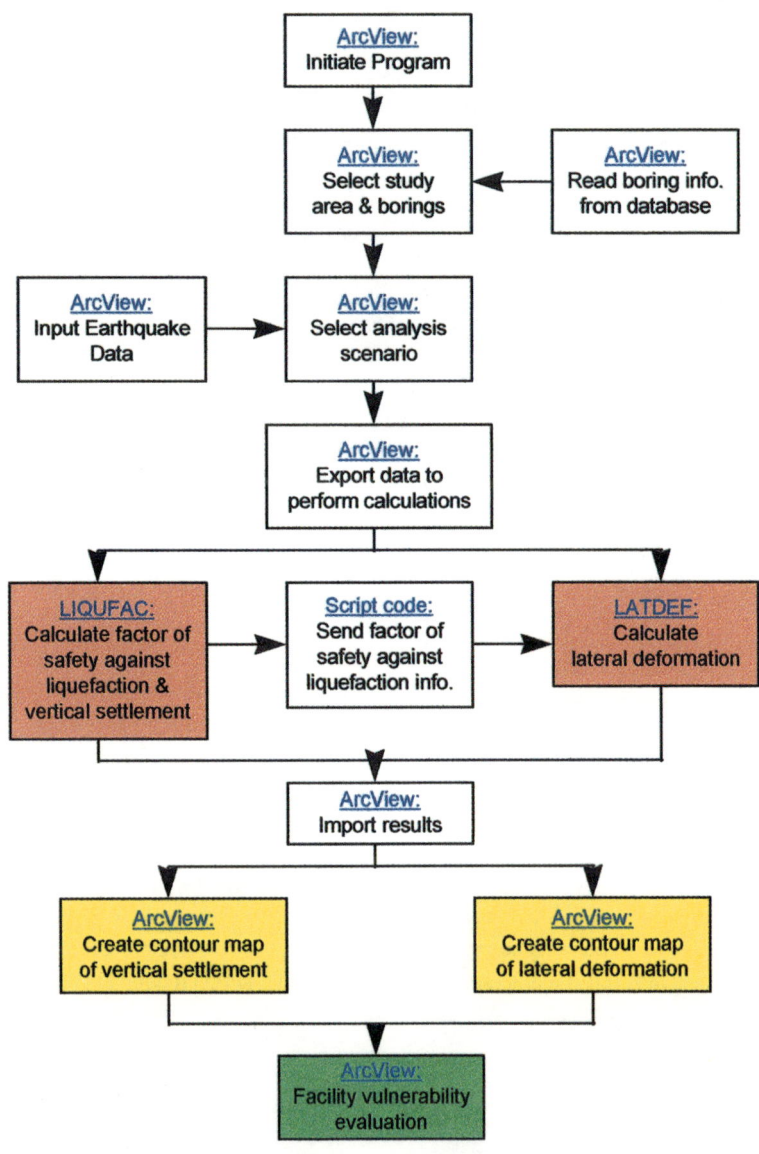

Figure 6 - Integrated Spatial Liquefaction Potential Mapping Procedure

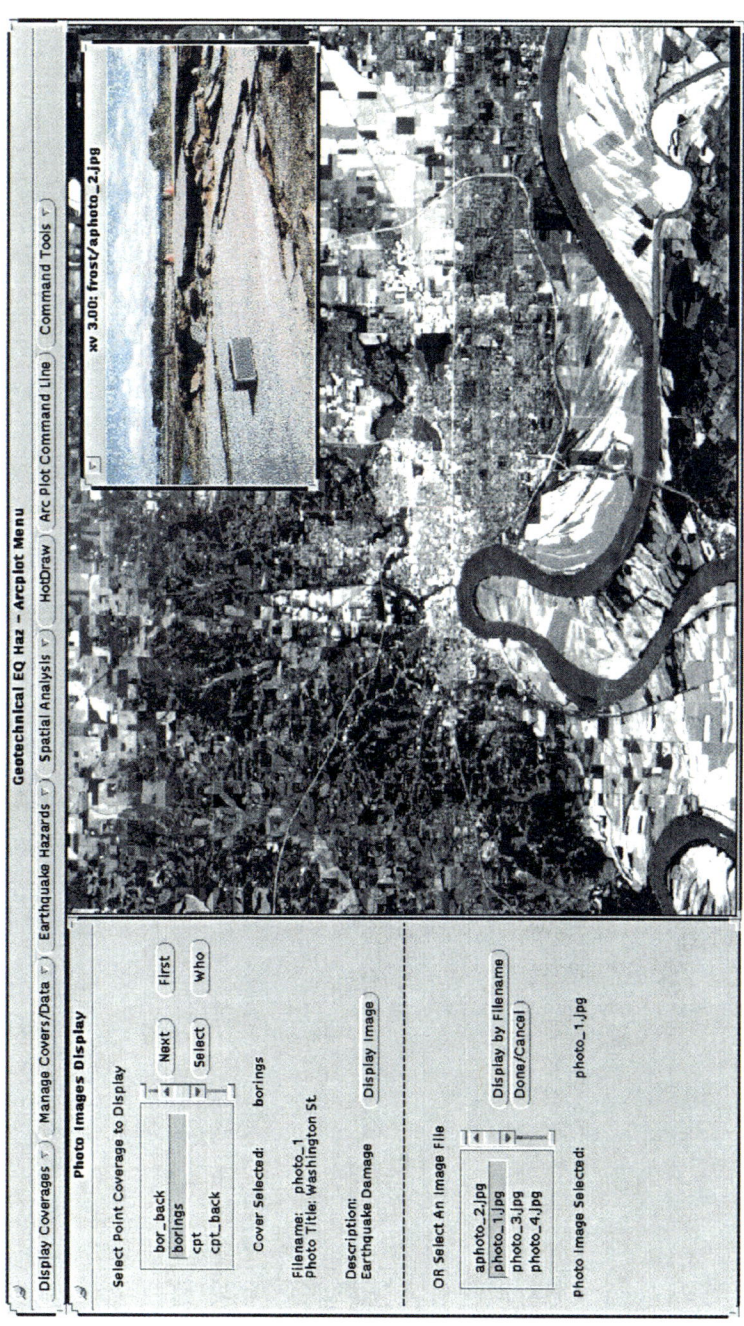

Figure 7   Combined Image Showing Simultaneous Display of Multiple Data Types

References

Bartlett, S. F., and Youd, T. L., (1992). "Empirical analysis of horizontal ground displacement generated by liquefaction-induced lateral spread." National Center for Earthquake Engineering Research, Technical Report NCEER-92-0021.

Castro, G., (1987). "On the behavior of soil during earthquake-liquefaction," Developments in Geotechnical Engineering 42, Soil Dynamics and Liquefaction, Edited by Cakmak, A. S., Department of Civil Engineering, Princeton University, Princeton, N.J., pp. 169-204.

Erten, D., and Maher, M.H., (1995). "Liquefaction Potential of Silty Soils", Proceedings of Seventh International Conference on Soil Dynamics and Earthquake Engineering, Computational Mechanics Publications, A.S. Cakmak and C.A. Brebbia (editors), pp. 163-171.

Evans, M.D., and Zhou, S., (1995). "Liquefaction Behavior of Sand-Gravel Composites", *Journal of Geotechnical Engineering*, ASCE, Vol. 121, No. 3, pp. 287-298.

Glaser, S.D., and Chung, R.M., (1995). "Estimation of Liquefaction Potential by In Situ Methods", *Earthquake Spectra*, EERI, Vol. 11, No. 3, pp. 431-455.

Ho, C.L., and Kavazanjian, E., (1991). "Reduction Factor for Liquefaction Potential Analysis", *Soil Dynamics and Earthquake Engineering*, Computational Mechanics Publications, Vol. 9, No. 6, pp. 313-322.

Iwasaki, T., Tatsuoka, F., Tokida, K., and Yasuda, S. (1978). "Estimation procedure of liquefaction potential and its application to earthquake resistance design." U.S. - Japan Panel on Wind and Seismic Effects, UJNR.

King, S.A., Kiremidjian, A.S., Law, K.H., and Basoz, N.I., (1995). "Earthquake Damage and Loss Estimation Through Geographic Information Systems", Proceedings of Fifth International Conference on Seismic Zonation, Vol. 1, pp. 265-272.

Luna, R., and Frost, J. D., (1995). Liquefaction Evaluation Using a Spatial Analysis System, School of Civil and Environmental Engineering, Georgia Institute of Technology, Report No. GIT-CEE/GEO-95-3, 202 pp.

Mayne, P.W., (1994). "Development of Piezo-Vibro Cone for Evaluating Soil Liquefaction Potential", Summary Report of Workshop on Advancing Technologies for Cone Penetration Testing for Geotechnical and Geoenvironmental Site Characterization, U.S. Army Research Office, J.J. Bowders and D.E. Daniel (editors).

Muraleetharan, K., Martin, G.R., and Heckma, L.K., (1991). "Recent Advances in Cone Penetrometer measurements and its Application to Liquefaction Susceptibility", Geotechnical Special Publication No. 29, Recent Advances in Instrumentation, Data Acquisition and Testing in Soil Dynamics, Bhatia and Blaney (editors), pp. 33-48.

Olsen, R.S., and Koester, J.P. (1995). "Prediction of Liquefaction Resistance Using the CPT", Proceedings of CPT '95, International Symposium on Cone Penetration Testing, Swedish Geotechnical Society, Vol. 2, pp. 251-255.

O'Rourke, T. D., and Pease, J. W. (1997). "Mapping liquefiable layer thickness for seismic hazard assessment." *Journal of Geotechnical Engineering*, ASCE, Vol. 123, No. 1, pp. 46-56.

Reyna, F.A.M., and Chameau, J.L., (1991). "Dilatometer Based Liquefaction Potential of Sites in the Imperial Valley", Proceedings of Second International Conference on Recent Advances in Geotechnical Earthquake Engineering and Soil Dynamics, Vol. 1, pp. 385-392.

Robertson, P. K., and Campanella, R. G. (1985). "Liquefaction potential in sands using the CPT." *Journal of Geotechnical Engineering*, ASCE, Vol. 111, No. 3, pp. 384-403.

Rockaway, T.D., and Frost, J.D., (1997). "Geotechnical Earthquake Hazard Assessment of the Evansville, Indiana Area", School of Civil and Environmental Engineering, Georgia Institute of Technology, Report No. GIT-CEE/GEO-97-5, 205 pp.

Seed, H. B., and Idriss, I. M. (1971). "Simplified procedure for evaluating soil liquefaction potential." *Journal of Geotechnical Engineering Division*, ASCE, Vol. 97, No. 9, pp. 1249-1273.

Seed, H.B., Tokimatsu, K., Harder, L.F., and Chung, R.M., (1985). "Influence of SPT Procedures on Soil Liquefaction Resistance Evaluations", *Journal of the Geotechnical Engineering Division*, ASCE, Vol. 111, No. 12, pp. 1425-1445.

Seed H.B., and DeAlba, P., (1986). "Use of SPT and CPT Tests for Evaluating the Liquefaction Resistance of Sands", ASCE Geotechnical Special Publication No. 6, Use of In Situ Tests in Geotechnical Engineering, pp. 281-302.

Singh, S., (1994). "Liquefaction Characteristics of Silts", Geotechnical Special Publication No. 44, Ground Failures under Seismic Conditions, pp. 105-116.

Suzuki, Y., Taya, Y., Tokimatsu, K., Kubota, Y., and Koyamada, K., (1995). "Field Correlation of Soil Liquefaction Based on CPT Data", Proceedings of CPT '95, International Symposium on Cone Penetration Testing, Swedish Geotechnical Society, Vol. 2, pp. 583-588.

Tokimatsu, K., and Yoshimi, Y., (1983). "Empirical Correlation of Soil Liquefaction Based on SPT N-Value and Fines Content", *Soils and Foundations*, Japanese Society of Soil Mechanics and Foundation Engineering, Vol. 23, No. 4, pp. 56-74.

Vaid, Y.P., and Thomas, J., (1995). "Liquefaction and Post-Liquefaction Behavior of Sand", *Journal of Geotechnical Engineering*, ASCE, Vol. 121, No. 2, pp. 163-173.

Youd, T. L., and Perkins, D. M. (1978). "Mapping of liquefaction induced ground failure potential." *Journal of the Geotechnical Engineering Division*, ASCE, Vol. 104, No. 4, pp. 433-447.

Youd, T. L., and Perkins, D. M. (1987). "Mapping of Liquefaction Severity Index." *Journal of Geotechnical Engineering*, ASCE, Vol. 113, No. 11, pp. 1374-1392.

DETERMINISTIC ZONATION OF SEISMIC SLOPE INSTABILITY: AN APPLICATION OF GIS

Carlton L. Ho[1], M. ASCE and Scott B. Miles[2]

Abstract

The state of the art in geographical information system (GIS) based seismic slope instability zonation is discussed with a review of seismic slope stability analysis. Regional landslide hazard assessment approaches are described with a focus on seismically induced landslides. These approaches include statistical and deterministic methods. Statistical techniques use weighting factors or multivariate regression models trained with causative parameters to describe potential instabilities. Based on physical models, deterministic approaches use pseudo-static or dynamic slope stability analysis to assess earthquake triggered landslide hazard. The use of GIS with each hazard zonation approach is illustrated through several case studies.

Introduction

Historically, seismically induced landslides have posed a great threat to both life and property. Keefer (1984) observed that earthquakes of moderate to high magnitude can cause landslides over an area as great as 500,000 km² for M = 9.2. This illustrates the need to assess the potential for seismically induced landslides on a medium (~1:25,000 - 1:50,000) to large (~1:5,000 - 1:10,000) scale. The static stability of a slope is influenced by geology, topography, hydrology, and land use. In addition to the static factors, seismic slope stability is influenced by earthquake magnitude, duration, frequency content, soil response, and source-to-site distance (Keefer, 1984; Sharma, 1996). Because of the extensive amount of required data, conventional seismic slope instability hazard zonation requires an inordinate amount of time to interpret and analyze data. The advent of geographic information systems (GIS) has provided a tool to effectively manage data interpretation and analysis required for medium to large scale seismic slope instability zonation.

This report reviews the state of the art in mapping seismically induced landslide hazard using GIS. Several popular methods for analyzing seismic slope stability are reviewed. Attention is paid to method limitations and application with GIS. A review of methods for general and seismic slope instability zonation and their application to GIS follows. Case studies are presented and discussed with respect to seismically induced landslide hazard zonation. Finally, ongoing research towards

---

[1] Assoc. Prof., Dept. of Civ. and Env. Engrg., Univ. of Mass., Amherst, MA 01003.
[2] Grad. Asst., Dept. of Civ. and Env. Engrg., Univ. of Mass., Amherst, MA 01003.

integrating conventional Newmark's displacement method (1965) within a GIS is discussed.

Seismic Slope Stability Analysis

Inertial instabilities are the focus of this report; as such, only methods for analyzing inertial seismic stability are reviewed. Methods of interest include pseudo-static analysis, Newmark's displacement method (1965), and simplified approaches to Newmark's method. Dynamic finite element analysis is not reviewed here because it is computationally impractical and the amount and quality of data required makes it cost prohibitive for medium and large scale hazard zonation.

Pseudo-Static Analysis

Pseudo-static analysis is a limit equilibrium method similar to static analysis. The primary difference between the two methods is the addition of constant horizontal and vertical forces that represent the inertial effects of an earthquake. The pseudo-static forces are assumed to be proportional to the weight of the slope by a seismic coefficient in the horizontal and vertical directions. The result of a pseudo-static analysis is expressed as a factor of safety against failure.

The inability of pseudo-static analysis to predict natural and man-made slope behavior during earthquakes has been clearly recognized. Any exceedance of the critical acceleration, no matter how brief, is considered to be failure. This definition relays little information as to the overall performance of a slope. Pseudo-static analysis is also considered to be largely unreliable because of the difficulty associated with determining appropriate seismic coefficients (Sharma, 1996).

The relative ease in which limit equilibrium analyses can be performed within a GIS has been well demonstrated (Terlien et al., 1993; Luzi and Pergalani, 1996; Herbel, 1994; Ho et al., submitted). Assuming infinite slope failure, pseudo-static analysis is computationally simple and straightforward to integrate with GIS using a single equation for determining safety factors.

Newmark's Displacement Method

Newmark (1965) observed that the transient effects of dynamic earthquake forces may result in permanent slope deformation that precedes complete failure. Each instant that the critical acceleration of a slope is exceeded equilibrium is not satisfied and, thus, the slope is accelerated. In this way, Newmark modeled a slope as a rigid friction block on an inclined plane. The method requires the determination of the critical acceleration for the given slope. This can be accomplished with a pseudo-static analysis with a unit factor of safety. Total displacement of a slope is calculated as the cumulative displacement of a friction block, having the same critical acceleration as the modeled slope, subjected to an input ground motion. Newmark's method provides information (permanent displacement) that is directly related to the input ground motion that can be judged as to the relative significance. Wilson and Keefer (1983) have demonstrated the relative accuracy of Newmark's method.

The sliding block analogy assumes that 1) the given slope is rigid and perfectly

88

plastic, 2) a well-defined planar slip surface exists, 3) there is negligible loss of shear strength during shaking, and 4) displacements occur only if dynamic stresses exceed shear resistance. Each assumption may only be valid for certain conditions and does not represent general soil behavior. The greatest difficulty associated with Newmark's method is the selection of a representative input accelerogram. Alternatives for input motions include selection of representative acceleration records, modification of existing time histories, or generation of artificial accelerograms. The soil response (amplification or deamplification of motion) also requires consideration. Using a finite element program or a 1-D solution (e.g. SHAKE) and input bedrock motions, one can estimate the average motions throughout the soil. A simple alternative is to apply an appropriate scaling factor to the bedrock motions.

In order to integrate Newmark's method with a GIS it necessary to develop a numerical scheme for calculating the cumulative displacement of each slope in the subject area. An algorithm for determining displacements can be programmed using either an internal or external programming language. Sharma (1996) and Jibson (1993) have written simple programs in FORTRAN and BASIC, respectively, based on the algorithm developed by Wilson and Keefer (1983). This algorithm double integrates those parts of the input accelerogram that exceed the critical acceleration. Computer resources and processing time constitute possible limitations to this approach.

Simplified Approaches to Newmark's Method

Several studies have been carried out to develop a simplified approach for obtaining Newmark displacements. The use of these approaches circumvents the necessity of selecting a representative accelerogram but still require knowledge of critical acceleration and various other parameters. Yegian et al. (1988) derived closed-form solutions for displacements caused by periodic rectangular, triangular and sinusoidal input motions. The solutions are presented as a chart of normalized displacement and critical acceleration ratio. Franklin and Chang (1977) double integrated 354 acceleration time histories to determine slope displacements in accordance with Newmark's method. The results are presented as curves of mean displacement, mean + σ, and a reasonable upper bound. Ambraseys and Menu (1988) developed multivariate regression equations, based on actual ground motions that express Newmark displacement, DN (in centimeters), as a function of critical acceleration, ac, and peak ground acceleration a_{max}. The following equation yields Newmark displacement if up-slope movement is not considered.

$$\log D_N = 0.90 + \log\left[\left(1 - \frac{a_c}{a_{max}}\right)^{2.53}\left(\frac{a_c}{a_{max}}\right)^{-1.09}\right] \qquad \sigma_{\log D_N} = 0.30 \qquad (1)$$

Yegian et al. (1991) considered frequency content and earthquake duration in developing the following equation based on the time histories of Franklin and Chang

(1977).

$$\log D_N^* = \log\left[\frac{D_N g}{a_{max} N_{eq} T^2}\right] = 0.22 - 10.12\frac{a_c}{a_{max}} + 16.38\left(\frac{a_c}{a_{max}}\right)^2 - 11.48\left(\frac{a_c}{a_{max}}\right)^3, \sigma_{\log D_N} = 0.45 \quad (2)$$

where N_{eq} is equivalent cycles and T is the predominant period of the input motion. Jibson (1993) used Arias intensity, I_a, rather than peak ground acceleration to better characterize the damaging effects of ground motion. The regression equation was developed using eleven acceleration time histories, ten from California, having Arias intensities as great as 10 $^m/_s$.

$$\log D_N = 1.460\log I_a - 6.642 a_c + 1.546 \qquad\qquad \sigma_{\log D_N} = 0.409 \qquad\qquad (3)$$

Makdisi and Seed (1978) based their method on Newmark's approach for calculating permanent displacements but extended the method to evaluate the dynamic response of soil. The procedure is iterative, estimating the shear strain within the soil, calculating the three modal periods for the slope, and determining the crest acceleration. A range of potential displacements is taken from a chart of critical acceleration ratio and displacement. A numerical example of the method is presented in Sharma (1996).

In general, the simplified approaches do not adequately characterize the input ground motion and, with the exception of Makdisi and Seed (1978), do not account for soil response. The majority of the approaches were developed as a function of critical acceleration ratio. Jibson (1993) states that there is no universal relationship between critical acceleration ratio and Newmark displacement. It is also commonly accepted that peak ground acceleration does not correlate well with earthquake damage. The model developed by Jibson using Arias intensity – shown to correlate better with earthquake damage – was trained with a limited number of input time histories that are specific to a geographic region. Application of the equation to other regions, such as Eastern North America, may lead to an underestimation of Newmark displacements (Jibson and Keefer, 1993). The equation also may not be reliable for Arias intensities greater than 10 $^m/_s$. The method developed by Makdisi and Seed (1978) accounts for soil response in a straight-forward but time consuming manner. The method, based on many simplifying assumptions may result in conservative values for displacement.

For GIS, the regression equation approaches are simple to implement – adding an attenuation relationship for determining peak ground acceleration or Arias intensity. The data required to perform a medium to large scale analysis is the same as for the pseudo-static approach in addition to knowledge of the causative fault location. Herbel (1994) and Ho et al. (submitted) employed the equation by Jibson (1993) to predict Newmark displacements for the East Bay Hills, California (1:24,000). The benefit of Makdisi and Seed (1978) is outweighed by the work required to implement the procedure within a GIS.

GIS in Landslide Hazard Zonation

Soeters and van Westen (1996) describe the following advantages to using GIS for landslide hazard zonation:

1. A much larger variety of hazard analysis techniques become attainable. Because of the speed of calculation, complex techniques requiring a large number of map overlays and table calculations become feasible.

2. It is possible to improve models by evaluating their results and adjusting the input variables. Users can achieve the optimum results by a process of trial and error, running the models several times, whereas it is difficult to use these models even once in the conventional manner. Therefore, more accurate results can be expected.

3. In the course of a landslide hazard assessment project, the input maps derived from field observations can be updated rapidly, where new data are collected. Also, after completion of the project, others can use the data in an effective manner.

Because of the volumes of data required to assess medium and large scale landslide hazard, data entry and verification are the greatest limitations of GIS applications. GIS methods in stability assessment require the design and construction of a database containing relevant information such as geology, topography, and hydrology among others. Some degree of data analysis – depending on zonation approach – is performed and the subsequent results are presented. The popular data structure for performing landslide hazard analysis is raster (pixel) based. One reason for this are that DEMs are stored as grids and require little manipulation for use with a raster based analysis. Raster based formats may not adequately represent morphological units. Vector based data structures permit better representation of morphology and provide a convenient means for storing attribute data. Presentation of analysis results can vary from one map showing temporal probabilities to a suite of maps depicting relative performance in various conditions. The ability of GIS to perform and present several variations of an analysis makes the output of a suite of maps the logical choice in characterizing landslide hazard.

The approach to landslide hazard zonation can be grouped into two categories: the earth science approach and the engineering approach (Luzi and Pergalani, 1996). The earth science approach can be further divided into the heuristic approach and the statistical approach. In reviewing the two approaches to landslide hazard zonation, several case studies are presented.

Earth Science Approach

The earth science approach assesses the potential for landslides through mapping of a few or many causative factors. This approach is convenient for applying to regional and medium scales where large amounts of accurate data are not readily available. Therefore, the disadvantage of such an approach is the low level of accuracy. Whether hazard is assessed primarily in the field or through statistical means indicates a heuristic or statistical approach, respectively.

The use of expert knowledge of regional causative conditions to map landslide hazard from direct field work or with photo interpretation is called heuristic analysis. Typically, there are no set rules for determination of potential instability. The approach may vary with each application and possibly with each slope in the region. With this technique, GIS is used only for plotting and data storage. If hazard is not determined directly in the field, qualitative weights may be assigned to each causative parameter and the overall hazard for each map unit determined as the sum of the assigned weights.

A heuristic approach for seismic instability zonation must consider factors such as magnitude, duration, and source-to-site distance. While it is possible to determine source-to-site distances with some degree of accuracy, other earthquake parameters can not be adequately assessed through field observation and expert knowledge. Therefore, seismic instability zonation with a heuristic approach may be unreliable even for relative comparison. The complex mechanisms of seismically induced landslides call for more reliable statistical or deterministic methods.

In the past, statistical approaches to slope instability zonation have been used extensively. Statistical approaches typically rely on the assumption that the past will predict the future. Relative hazard is assessed by establishing statistical correlations between past incidences of landslides and causative parameters. The primary steps in any statistical approach are determining areal landslide density and mapping the identified factors that directly or indirectly control slope stability.

If a statistical weighting approach is taken, landslide density is determined with respect to each causative parameter. This can be accomplished by overlaying individual parameter maps onto the landslide density map. Weights are subsequently assigned to each parameter map based on the incidences of landslides within each particular parameter map (Figure 1). Weighting factors are typically a function of the landslide density corresponding to each parameter. The relative landslide hazard is determined by summing the weights from each parameter map. Statistical weighting introduces subjectivity in the selection of causative parameter and, possibly, in the assignment of weights. Another shortcoming of the approach is the difficulty in accounting for interaction between identified parameters.

A good example of a statistical weighting approach is found in a study by Gupta and Joshi (1990) of the 3,135 km^2 Ramganga catchment (India). Landslide density, lithology, land use, distance from major tectonic shear zone, and slope aspect were used to calculate a landslide nominal risk factor (LNRF) for each identified input parameter. LNRF was defined as landslide incidence in a particular parameter (e.g. land use) sub-category (e.g. agriculture, forest, barren) divided by the average incidence in each sub-category. LNRF was used to integrate the various factors of different units and scales. Weights were assigned to each parameter map pixel based on the calculated LNRF value. The weights were summed to assess whether each

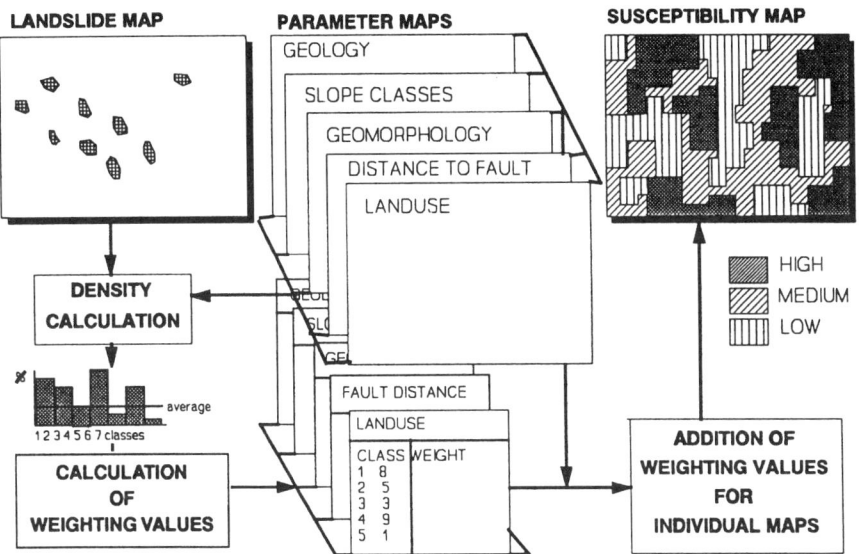

Figure 1. Use of GIS with statistical weighting approach.

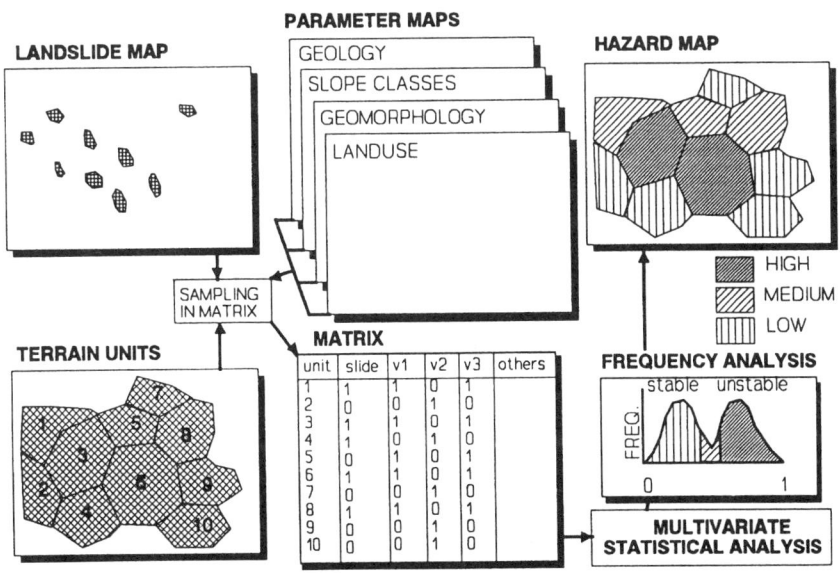

Figure 2. Use of GIS with statistical regression approach.

93

pixel has low, medium, or high risk. The hazard associated with earthquakes was only indirectly considered with the parameter distance from major tectonic shear zone (not to be confused with distance from fault).

An alternative technique to weighting factors using statistical methods is to calibrate a multivariate regression model of landslide probability as a function of identified causative factors. This approach accounts for interaction between factors that directly and indirectly control stability (Wang and Unwin, 1992). Such techniques typically incorporate an extensive data set of causative parameters. Thus, identifying and mapping the parameters comprise a significant step of the analysis. Regression models are calibrated by creating a coverage describing the stability of each morphological unit with respect to controlling factors. The coverage is created using map overlay functions of GIS (Figure 2). If the analysis is raster based, a means of integrating the data layers is required. Attribute tables are used with vector based analysis for storing information. The calibrated model can then be used to determine the probability of instability for other areas. Because the model is trained with an original subject area results may not be reliable for areas with dissimilar conditions. Another limitation attributed to the multivariate approach is the inordinate amount of causative parameters, and thus, data that is needed to obtain a satisfactory model.

Carrara et al. (1991) calibrated a multivariate regression model with a 60 km^2 area in the Tescio river basin (Italy) using discriminant analysis. In order to perform the analysis, each slope unit was associated with 40 morphological, geological and vegetation attributes. Of the attributes, slope gradient, height and length were not included. Carrara states that the model cannot be readily extended to other areas with different causative factors. Wang and Unwin (1992) compared a multivariate log-linear model, trained on a 225 km^2 area in the loess fields of China, with sieve mapping (map overlay) and weighting factor techniques. The study found that the log-linear regression model is superior to the more simplistic approaches in that each causative factor is weighted in an optimal sense based on past information. Chung et al. (1993) introduced a regression technique that can be directly applied to vector based GIS without rasterization. The model's ability to predict slope instability was evaluated by pretending the present was 1960. Based on pre-1960 information, predictions were favorably compared with post-1960 landslide distribution.

Engineering Approach

The engineering approach is a deterministic one. The use of physical models in conjunction with GIS to predict slope performance for medium to large scales is becoming increasingly popular although the history is quite brief. Deterministic approaches have an advantage over statistical ones by incorporating geotechnical engineering analysis. Geotechnical slope stability analysis is based on physical properties and mechanical behavior and can be extended to evaluate seismic slope stability. The data required for landslide hazard zonation using an engineering approach are those required for each particular slope stability model. These data include material properties (unit weight and shear strength), depth to failure, slope geometry, and depth to groundwater. For seismically induced landslide hazard, predominant fault location and ground motion data are also required. The need for quality data is the greatest disadvantage of the engineering approach. For medium scale analysis it may not be economically feasible to perform extensive lab and field testing. Alternatively, a small number of laboratory and field tests can be performed in order to extrapolate parameters to untested areas. The use of existing data from sources such as geologic maps, soil surveys or previous private or public projects constitutes another option. An additional consideration, with respect to data quality, is the accuracy of topographic data. The necessity for quality data for use in deterministic models points towards a need for multidisciplinary GIS (e.g. environmental, transportation, and geotechnical) so that the costs of data collection and data capture can be shared.

After data input and verification, slope stability can be analyzed in one of two ways. Equations derived for determining a safety factor against failure can be applied. The infinite slope model is the easiest and most popular method, although, a method of slices has been used within a GIS (Luzi and Pergalani, 1996). The alternative is to export the data for each pixel (raster based), morphological unit (vector based), or selected profile for use in an external slope stability program. This approach does not take full advantage of the benefits provided by GIS. To date, there has been no attempt at integrating a slope stability program inside of a GIS. To assess seismic slope stability, either pseudo-static or dynamic (e.g. Newmark's method) analysis can be performed. The presentation of deterministic results can take many forms. A map showing temporal probability of a factor of safety less than one can be created. Alternatively, a map (or a suite of maps) showing slope performance under certain conditions can be created. Figure 3 illustrates possible approaches to deterministic landslide zonation and its presentation using GIS.

Research by Terlien et al. (1993) applied a raster based GIS in deterministic landslide hazard zonation. The study focused on the use of deterministic hydrological models for accurate assessment of groundwater levels and pore pressure for use with external and internal analyses of static and pseudo-static slope stability. Three applications of deterministic landslide hazard zonation were presented. An application from Costa Rica utilized a one-dimensional external hydrological model to determine the height of perched water tables for various soil types and different rain storms. After obtaining shear strength data from direct shear tests, static stability was evaluated using the infinite slope model. In the first application from Columbia, an external two-dimensional hydrologic model was used to calculate the minimum groundwater depth for a 20 year return period. An internal pseudo-static analysis on data extracted from geologic maps assessed seismic instability hazard. The study made no reference to the determination of seismic coefficients. Terlien et al. (1993) cited inadequate data and limitations of their particular GIS software (ILWIS) as reasons for not performing dynamic slope stability analysis. An internal three-dimensional hydrological model was used to simulate changes in the groundwater table over a rainy season for the final application from Columbia. The groundwater maps were exported for analysis in a static slope stability program.

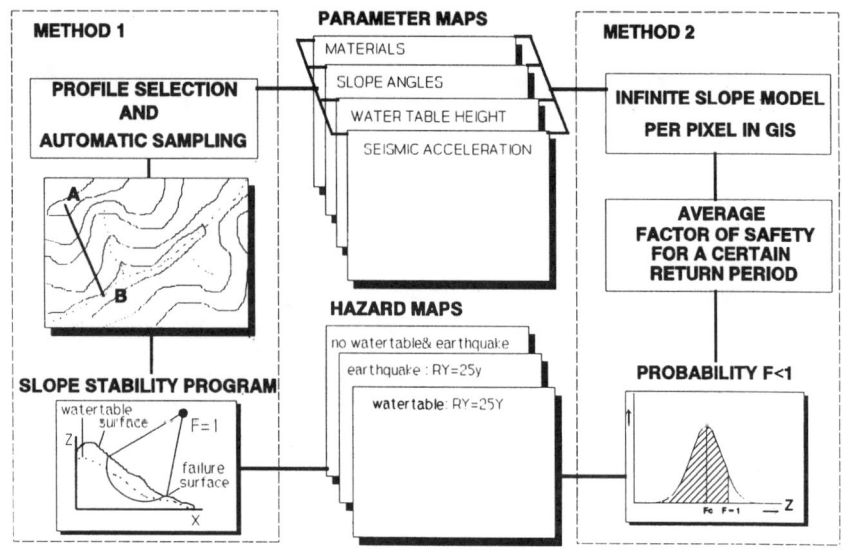

Figure 3. Use of GIS with deterministic approach.

Hazard analysis results for each application were used to obtain maps depicting the spatial distribution of safety factors and the probability of failure.

For the north-east section of the 1:50,000 Fabriano (Italy) geologic map, Luzi and Pergalani (1996) assessed areal slope stability using static, pseudo-static, and dynamic analysis with a raster based GIS. Pertinent soils data were determined from geologic maps, limited laboratory testing, and existing data supplied by private companies. The soil was assumed cohesionless so that factor of safety is not a function of failure depth. Static and pseudo-static analysis were performed on each pixel using the popular infinite slope model. Equations for performing a method of slices analysis using individual pixels as slices were developed and used for static and pseudo-static conditions. The influence of surrounding pixels on each particular cell was accounted for through neighborhood analysis functions of GIS. Maps of static safety factors were created for combinations of maximum and minimum geotechnical parameters with dry and saturated conditions. Pseudo-static maps of critical seismic coefficients were created for the same parameter combinations. Seismically induced displacements were calculated with a dynamic analysis. Considering each pixel as independent blocks sliding on inclined planes, Newmark's (1965) displacement method was adapted for use within the GIS. The actual approach to the application of Newmark's method, whether conventional or simplified, was not specified. Results from the application of two different accelerograms during dry conditions were shown as maps of predicted displacements. No seismic source was defined for the dynamic analysis and, thus, the ground motions were not attenuated throughout the region. Luzi and Pergalani (1996) cited the low accuracy of geotechnical and topographic data as the greatest limitation of their methodology.

A vector based GIS was used in developing a seismically-induced displacement map for the East Bay Hills region of the 1:24,000 Oakland East quadrangle (Herbel, 1994; Ho, et al., submitted). The hazard map presented zones of relative landslide displacements determined with a simplified approach to Newmark's displacement method. Soil parameters were estimated from a geologic map of the quadrangle and expert knowledge. The geologic map was digitized and attribute data (e.g. shear strength and unit weight) assigned to each morphological unit. For each morphological unit, a static factor of safety under dry conditions and constant depth was calculated using the infinite slope model. With this information, critical accelerations were calculated with a relationship by Newmark (1965). Ground motion was characterized for a $M = 7$ design earthquake with Arias intensity using an attenuation relationship. Distance between morphological units and the Hayward fault trace were determined using spatial analysis functions of GIS. Landslide displacements were calculated using the multivariate regression equation by Jibson (1993). The results of the hazard analysis were shown as a single map of predicted displacements.

Use of Conventional Newmark's Method with GIS

Previous applications of seismic slope stability analysis to GIS have utilized simplistic pseudo-static methods or limited approaches to Newmark's displacement method. Pseudo-static landslide zonation yields little meaningful information as to slope performance; presenting either a factor of safety or some critical acceleration value. In response to the shortcoming of pseudo-static approaches, simplified approaches to Newmark's displacement method have been integrated with GIS for predicting areal slope performance. These approaches are limited with respect to geographic application and may suffer from inadequate characterization of local ground motion. Current work by the authors, motivated by these observations, integrates the conventional approach to Newmark's method with a GIS for mapping potential displacements resulting from a design earthquake. This work addresses the shortcoming of pseudo-static methods while improving on the limitations of the previous approaches to Newmark's method.

The subject of the hazard analysis is the East Bay Hills area (30 km^2) which is located in the vicinity of Berkeley, California (Figure 4). The predominant seismic source in the coverage is considered to be the Hayward fault. The fault zone runs northwest to southeast near the foot of the East Bay Hills. Soils in the coverage consist primarily of silty sand and low-plasticity silts with areas of low-plasticity clay and artificial fill (Figure 4a). The general geology, going from southwest to northeast, includes formations of sandstone, shale, chert, basalt, and claystone (Figure 4b).

The first step in conducting the Newmark's analysis is determining the critical acceleration of each slope. This value is determined with a relationship by Newmark (1965) that expresses critical acceleration as a function of static factor of safety and thrust angle. The safety factor is calculated using the infinite slope model. Soil properties estimates have been obtained from soils surveys, a geologic map, and engineering judgement. Drained strength parameters are used to simplify data collection for the prototype GIS.

For the conventional Newmark's method, the next step is to double integrate those parts of an acceleration times history that exceed the critical acceleration. In order to attenuate ground motion and characterize future earthquakes, artificial accelerograms created with a shot-filtered white noise technique developed by Boore (1983) are used. A robust macro determines the displacement resulting from a design earthquake (M = 7) for each morphological unit in the coverage using the double integration algorithm. The program selects a specific time history for integration based on the source to site distance of each map unit. Because of the random nature of the time history generation the analysis should be performed several times to adequately describe the hazard.

Results of one realization of the hazard analysis, for dry and saturated conditions, are shown in Figure 5. After subsequent trials, the analysis results may be presented as suite of maps depicting probabilistic hazard for various seismic and geotechnical conditions. This approach to assessing seismic slope instability can be used as a part of general seismic risk (e.g. King, 1994) or seismically induced landslide risk (e.g. Ho et al., 1995) methodologies for GIS.

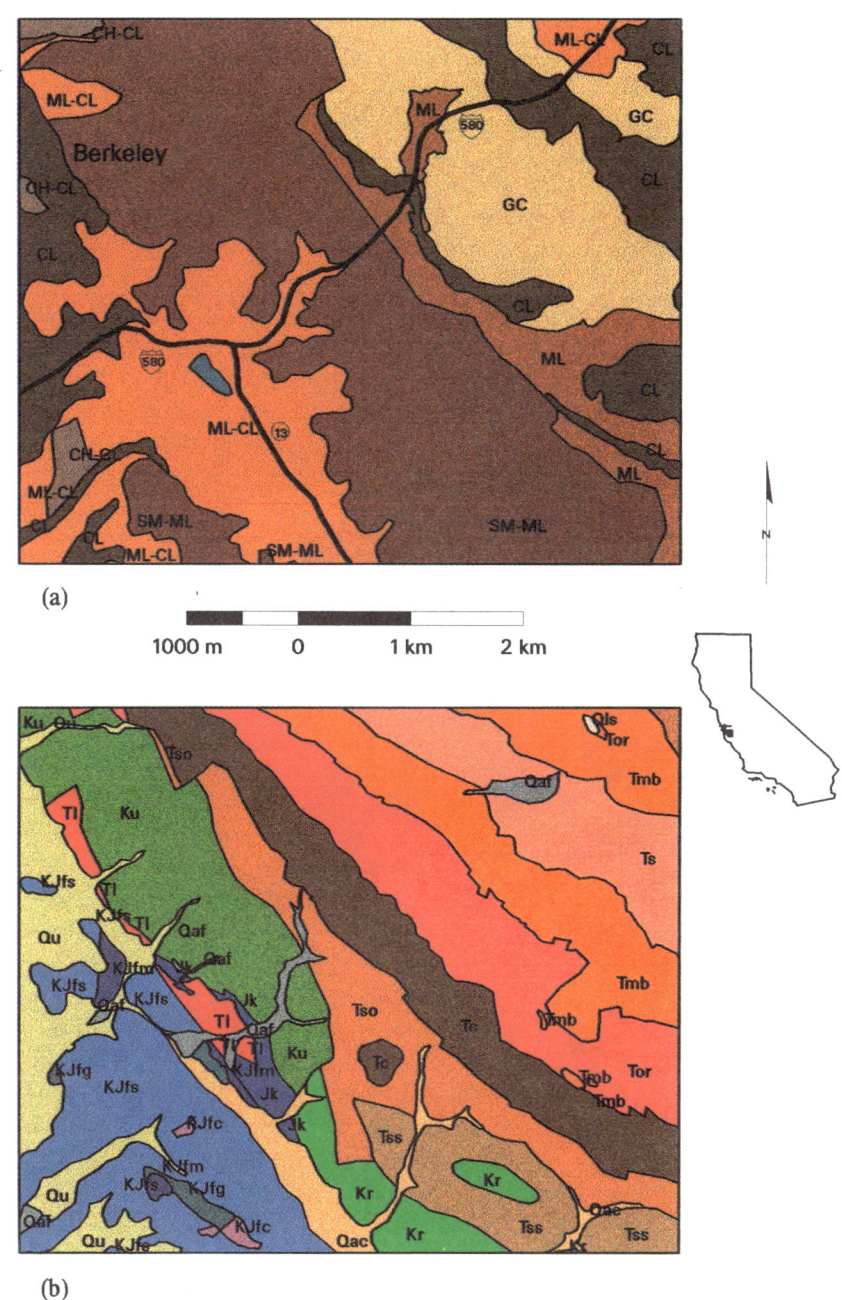

Figure 4. East Bay Hills, California. a) Regional soils b) Areal geology.

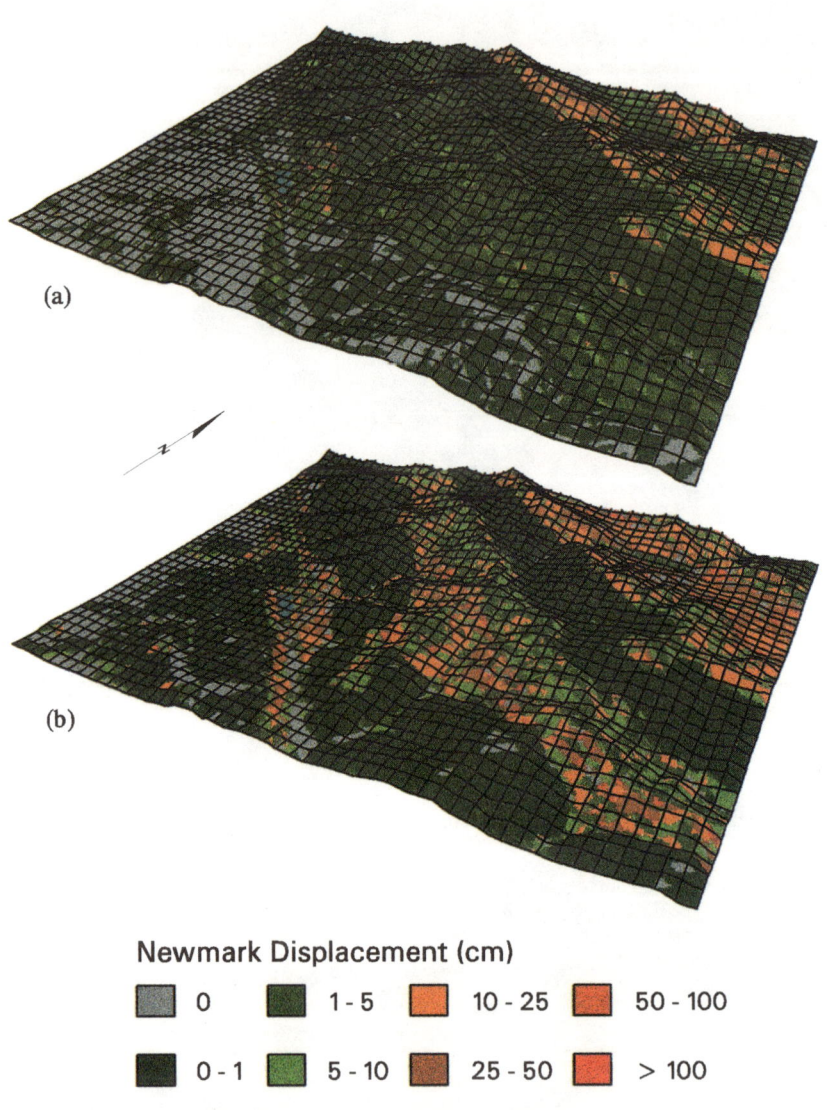

Newmark Displacement (cm)

⬜ 0	🟩 1 - 5	🟧 10 - 25	🟧 50 - 100
⬛ 0 - 1	🟩 5 - 10	🟫 25 - 50	🟥 > 100

Figure 5. Predicted Newmark displacement for an M = 7 earthquake on the Hayward fault. a) Dry conditions. b) Saturated conditions.

Conclusion

The union of GIS and deterministic slope stability analysis is a significant advancement in landslide hazard zonation. GIS effectively manages the large amounts of data required of geotechnical analysis. To date, the infinite slope model remains the most popular deterministic model for both static and seismic slope instability zonation. Approaches such as pixel-based method of slices and external use of existing slope stability programs are potential alternatives. The use of an existing slope stability program in a landslide hazard GIS is a logical next step.

Three general deterministic approaches have been used for seismically induced landslide hazard zonation. The pseudo-static approach requires the fewest data and is computationally simple, but fails to provide meaningful information on slope performance. The simplified approaches to Newmark's method provide estimates of permanent displacements with only a small increase in computational time or required data. For greater flexibility and ground motion characterization, the conventional Newmark's analysis can be applied to GIS. Pore pressure effects and soil amplifications should be addressed in future GISs for seismically induced landslide hazard assessment. Finally, information regarding the distribution of earthquake triggered landslides from events such as the Northridge earthquake provide a significant opportunity to verify and calibrate deterministic seismic slope instability predictions on a medium to large scale.

Bibliography

Ambraseys, N.N. and Menu, J.M., (1988) "Earthquake-Induced Ground Displacements", *Soil Dyn. and Earthquake Engrg.*, **16**, pp. 985 – 1006.

Boore, D.M., (1983) "Stochastic Simulation of High Frequency Ground Motions Based on Seismological Models of the Radiated Spectra", *BSSA*, **73**, 6, pp. 1865 - 1894.

Carrara, A., Cardinali, M., Detti, R., Guzzetti, F., Pasqui, V., and Reichenback, P., (1991) "GIS Techniques and Statistical Models in Evaluating Landslide Hazard", *Earth Surface Processes and Landforms*, **16**, pp. 427 - 445.

Chung, C.F., Fabbri, A.G., and Van Westen, C.J., (1993) "Multivariate Regression Analysis for Landslide Hazard Zonation", *GIS in Assessing Natural Hazards*, pp. 107 - 122.

Franklin, A.G. and Chang, R.K., (1977) "Permanent Displacements of Earth Embankments by Newmark's Sliding Block Analysis", *Rpt. 5, Misc. Paper S-71-17*, U.S. Army Corps Engr. W.E.S. Vicksburg, MI.

Gupta, R.P. and Joshi, B.C., (1990) "Landslide Hazard Zoning Using the GIS Approach -- A Case Study from the Ramganga Catchment, Himalayas", *Engrg Geo.*, **28**, pp. 119 - 131.

Herbel, W.A., (1994) "Development of a GIS Based Landslide Hazard Map", *M.S.C.E.*, Wash. St. Univ., Pullman, WA.

Ho, C.L., Herbel, W., Keefer, D.K. and Miles, S.B., (In Review) "Development of a GIS-based Landslide Hazard Map", *AEG*.

Ho, C.L., Hadj-Hamou, T.A., Nilsson, Michael, (1995) "GIS Based Zonation of Infrastructure Damage Related to Seismically Triggered Landslide Risk," proc. 5th Int. Conf. on Seismic Zonation, AFPS/EERI, Nice, Oct. 1995.

Jibson, R.W., (1993) "Predicting Earthquake-Induced Landslide Displacements Using Newmark's Sliding Block Analysis", *Trans. Res. Rec. 1411*, pp.1-15.

Jibson, R.W. and Keefer, D.K., (1993), "Analysis of the Seismic Origin of Landslides: Examples from the New Madrid Seismic Zone", *GSA. Bull.*, **105**, pp 521 - 536.

King, S.A. and Kiremidjian, A.S., (1994) "Regional Seismic Hazard and Risk Analysis Through Geographic Information Systems", *Ph.D.* Stanford University.

Luzi, Lucia and Pergalani, Floriana, (1996) "Applications of Statistical and GIS Techniques to Slope Instability Zonation (1:50,000 Fabriano geological map sheet)", *Soil Dyn. and Earthquake Engrg.*, **15**, 2., pp. 83 - 94.

Makdisi, F.I. and Seed, H.B., (1978) "Simplified Procedure for Estimating Dam and Embankment Earthquake Induced Deformations", *ASCE J. Geo. Engrg*, **104**, GT7, pp. 849 - 867.

Newmark, N.M., (1965) "Effects of Earthquakes on Dams and Embankments", *Geotechnique*, **15**, pp. 139-160.

Sharma, Sunil., (1996) *Chap 6: Slope Stability and Stabilization Methods*, Ed. Abramson, L.W., Lee, T.S., Sharma, S., and Royce, G.M. New York: Wiley.

Shu-Quiang, Wang and Unwin, D.J., (1992) "Modeling Landslide Distribution on Loess Soils in China: an Investigation", *Int. J. GIS*, **6**, 5, pp. 391 - 405.

Soeters, R. and van Westen, C.J., (1996) "Slope Instability Recognition, Analysis, and Zonation", *Trans. Res. Board Spec. Rep. 247: Landslides Investigation and Mitigation*, pp.129-177.

Terlien, M.T.J., Van Westen, C.J. and Van Asch, T.W.J., (1993) "Deterministic Modelling in GIS-Based Landslide Hazard Assessment", *GIS in Assessing Natural Hazards*, pp. 57 - 77.

Wilson, R.C. and Keefer, D.K., (1983) "Dynamic Analysis of a Slope Failure From the 6 August 1979 Coyote Lake, California, Earthquake", *BSSA*, **73**, 3, pp. 863 - 877.

Yegian, M.K., Marciano, E., and Ghahraman, V.G., (1991) "Earthquake-Induced Permanent Deformations: Probabilistic Approach", *ASCE J. Geo. Engrg.*, **17**, 1, pp. 35 – 50.

Yegian, M.K., Marciano, E., and Ghahraman, V.G., (1988) "Integrated Seismic Risk Analysis for Earth Dams", *Report 88-15*, Northeastern Univ, Boston.

Integration of Earthquake Hazards in GIS

Stephanie A. King[1]

Abstract

This paper presents a comparison among three regional earthquake loss estimation loss methodologies: ATC-13 (Applied Technology Council, 1985), ATC-36 (Applied Technology Council, in progress), and the standardized methodology prepared for the National Institute of Building Sciences (NIBS; Risk Management Solutions, 1995). The methods for computing damage to wood frame buildings (the most common type of residential structure in California) as a result of ground shaking, liquefaction, and landslide, as well as the combination of the damages, are discussed and compared. The models are implemented in a GIS for a case study region. Results of the comparison show that the damage due to ground shaking is very similar, but that the collateral effects and combined damage estimates are significantly different. The results are applicable to wood frame buildings subjected to a large scenario earthquake (a magnitude 8.0 on a close fault). Other building types and ground shaking distributions could produce different results.

Introduction

Earthquake damage to structures can be attributed to strong ground shaking as well as secondary or collateral earthquake-induced effects. These secondary effects include liquefaction, landslide, fault rupture, inundation, and fire-following earthquake. The combination or integration of the damage caused by ground shaking and secondary effects has been addressed in recent research projects as well as in several regional earthquake damage and loss estimation methodologies.

Kiremidjian (1992) and King and Kiremidjian (1994) have suggested a weighted-average approach to combining the effects of ground shaking and secondary hazards, recognizing that all of the secondary effects are not likely to occur simultaneously. This approach has been shown to be sensitive to the selection of the

---

[1] Associate Director, John A. Blume Earthquake Engineering Center, Stanford University, Stanford, California, 94305-4020

weighting factors on each hazard. The ATC-13 study includes a conservative approach to combining the various hazards for regional earthquake damage and loss estimation. In ATC-13, the mean damage factor is computed for ground shaking and for each individual hazard and then the damage factors are added together as a straight sum, bounded at 100%. The ATC-36 study is an improvement to the ATC-13 methodology in that the damage factors from ground shaking and each individual hazard are added together in an SRSS (square root of the sum of the squares) approach with estimates of the maximum and minimum bounds. The standardized earthquake loss estimation methodology prepared for NIBS treats the combination of ground shaking damage and secondary effects by adding the individual probabilities of being in or exceeding discrete damage states due to ground shaking and ground failure due to permanent ground deformation.

This paper presents a comparison among the earthquake damage combination methods presented in three regional earthquake damage and loss estimation methodologies, namely ATC-13, ATC-36, and NIBS. The individual hazard models and methods for estimating damage due to each hazard are also compared along with the three combination methodologies. The collateral hazards that are considered in this paper are limited to liquefaction and landslide. Other secondary effects such as inundation, fault rupture, and fire-following earthquake are either treated very similarly in the various methodologies or are considered too complicated for comparison in a paper of limited length.

Regional Earthquake Damage Estimation

Hazard and damage estimation are only one part of an overall earthquake damage and loss estimation methodology; the other components include inventory development and loss modeling. The hazard and damage methods described in this paper are for estimation on a regional basis and include many assumptions and simplifications, particularly with respect to the regional geologic data used in the modeling. The methods are not intended for site-specific analysis, as site investigations would yield more accurate geotechnical data that would be applicable for more rigorous modeling.

The models are regional in nature and involve the combination of several types of spatially distributed data, making them ideal for implementation in a geographic information system (GIS). The comparison of regional earthquake hazard and damage models is applied to a case study for the city of Palo Alto located in northern California. A scenario earthquake of magnitude 8.0 on the San Andreas fault is used for the comparison; however, the methods are applicable for other scenario events and probabilistic hazard analyses. For comparison purposes, the uncertainties in each model are ignored and damage is limited to structural and nonstructural damage (contents damage is not included in the analysis). Each hazard and damage model is briefly described, followed by a description of the damage integration methods and the results of the case study application.

Strong Ground Shaking

Structural damage due to earthquake shaking is estimated with motion-damage relationships. The relationships are typically based on at least one of three approaches: (1) consensus expert opinion, (2) analytical modeling, and (3) empirical data. A given classification of structural types is developed for the region, and for each class, the expected damage is estimated for various levels of input ground motion. In regional damage and loss estimation, an inventory of structures is compiled and classified according to structural type. The distribution of ground shaking in the region is estimated for scenario events or probabilistic hazard analysis, and the motion-damage relationships are applied to produce damage estimates for the structural inventory.

The ATC-13, ATC-36, and NIBS have different approaches to the estimation of damage due to ground shaking; however, the structural classifications are similar. The differences in damage estimation are in the ground motion parameter and the development of the motion-damage relationships. The ATC-13 and ATC-36 methods use Modified Mercalli Intensity (MMI) as the ground motion parameter and the motion-damage relationships are based on expert opinion. In the ATC methods, damage due to ground shaking is estimated as a mean damage factor, E[DF], the ratio of the damaged value to the replacement value of the structure. The NIBS method uses spectral acceleration (S_a) and spectral displacement (S_d) as the ground motion input parameters and the motion-damage relationships are based on a combination of analytical modeling and expert opinion. In the NIBS method, damage due to ground shaking is estimated as the probability of being in or exceeding five damage states that range from none to complete.

Liquefaction

Liquefaction is the loss of shear strength in loose saturated soil when subjected to earthquake shaking. The susceptibility of a region to earthquake-induced liquefaction is typically mapped into qualitative classes such as high, moderate, low, and very low based on geotechnical investigations of soil parameters including age, lithology and level of ground water table. Given the susceptibility map, the three regional earthquake loss estimation methods include different approaches for estimating liquefaction-induced damage. Only damage to surface facilities will be addressed in this paper.

In the ATC-13 methodology, the probability of liquefaction is estimated for every combination of input MMI level and soil deposit. Damage due to liquefaction is computed as the damage due to ground shaking multiplied by the probability of liquefaction for the given soil deposit and input MMI level, and then multiplied by a factor of 5. The ATC-36 method for estimation of liquefaction-induced damage is similar to the ATC-13 method in that it is also based on the ground shaking damage and the probability of liquefaction at the site, which is a function of the soil deposit and input MMI level. In the ATC-36 method, liquefaction damage is computed as the probability of liquefaction for the given soil deposit and input MMI level multiplied by the damage due to ground shaking if the ground shaking level were increased by one MMI level.

In the NIBS methodology, the probability of liquefaction is estimated as a function of the input PGA, the depth to groundwater, the earthquake magnitude, and the susceptibility of the soil deposit to liquefaction. Liquefaction damage is assumed to be a function of the displacement in terms of lateral spreading and displacement. Lateral spreading displacement is estimated as a function of the ratio between the input PGA and the PGA required to induce liquefaction for the given susceptibility class, while ground settlement is a function only of the susceptibility class. Liquefaction-induced damage is assumed to cause only extensive or complete building damage, i.e., buildings are assumed to be either undamaged or severely damaged due to permanent ground deformation. The liquefaction damage is estimated using curves that give the probability of having extensive or complete damage as a function of the amount of lateral spreading and settlement.

Landslide

The estimation of earthquake-induced landslide damage is similar to the estimation of earthquake-induced liquefaction damage. The susceptibility of a region to earthquake-induced landslide is also typically mapped into qualitative classes such as high, moderate, low, and very low based on geotechnical investigations of soil parameters including age, lithology and slope angle. Given the susceptibility map, the three regional earthquake loss estimation methods include different approaches for estimating landslide-induced damage.

In the ATC-13 methodology, six classes of slope stability are defined ranging from unstable to very stable. For each stability class, six slope failure damage states are defined with associated probabilities of occurrence for various input MMI levels. For a given MMI level and slope stability class, the landslide-induced damage is computed as the sum of the products of the damage factor associated with each slope stability damage state and the probability of that damage state occurring. In the ATC-36 methodology, landslide potential in a region is mapped into four classes, each having probabilities of occurrence for various levels of input MMI. Landslide damage is estimated using empirical relationships between land movement and landslide potential class, and between landslide-induced damage to a structure and land movement at the site. The expected damage factor due to landslide is found by multiplying the expected landslide damage by the probability of the landslide occurring, which is a function of the input MMI level and landslide potential class.

In the NIBS methodology, sites are classified into one of ten landslide susceptibility classes according to type of deposit and slope angle. The probability of landslide is given for each susceptibility class. The permanent ground displacement is estimated from a relationship between displacement and ground acceleration, the input PGA, and the number of earthquake cycles (a function of the earthquake magnitude). Similar to the estimation of liquefaction damage in the NIBS methodology, landslide-induced damage is assumed to induce only extensive or complete building damage, i.e., buildings are assumed to be either undamaged or severely damaged due to permanent ground deformation. The landslide damage is estimated using curves that give the probability of having extensive or complete damage as a function of the amount of permanent ground deformation.

<u>Integration of Earthquake Damage</u>

The ATC-13 methodology includes the most conservative method for combining the damage from ground shaking and collateral effects. The individual damage factors are added in a straight sum, and truncated to 100% if the sum is greater than 100%. Considering only the secondary effects of liquefaction and landslide the total expected damage factor for a given site would be computed as follows:

$$EDF[T] = EDF[S] + EDF[L] + EDF[LS] \tag{1}$$

where:

EDF[T] $\leq$ 100%
EDF[S] is the expected damage factor for ground shaking damage
EDF[L] is the expected damage factor for liquefaction damage
EDF[LS] is the expected damage factor for landslide damage

The ATC-36 methodology includes what is intended to be an improvement to the ATC-13 method for combining the damage from ground shaking and collateral effects. The most likely estimate of the total expected damage is computed as the SRSS (square root of the sum of the squares) of the individual damage factors. This approach was selected because sensitivity studies showed that a straight sum tended to overestimate the total damage, while a straight average tended to underestimate the damage. The SRSS method is almost a weighted average approach in that the higher damage factors contribute more in the SRSS damage computation due to the squaring of the terms. Considering only the secondary effects of liquefaction and landslide the total expected damage factor for a given site would be computed as follows:

$$EDF[T] = \{EDF[S]^2 + (EDF[L] - EDF[S])^2 + EDF[LS]^2\}^{1/2} \tag{2}$$

where:

EDF[T], EDF[S], EDF[L], and EDF[LS] are defined as in Equation 1

In the NIBS methodology, damage is given in terms of probabilities of having damage that is equal to or exceeds five damage states (none, slight, moderate, extensive, and complete). Damage due to ground failure or permanent ground displacement is assumed to either cause no damage or extensive damage. In order to combine the damage due to ground shaking with damage due to ground failure, it is assumed that the probability of having at least slight and moderate damage due to ground failure is the same as the probability of having at least extensive damage due to ground failure. The probability of having complete damage due to ground failure is assumed to be 20% of the probability of having at least extensive damage due to ground failure. Given these assumptions, the combined or total probabilities of damage the various damage states due to ground shaking and ground failure are as follows:

$$P_{COMB}[DS \geq S] = P_{GF}[DS \geq S] + P_{GS}[DS \geq S] - P_{GF}[DS \geq S] \times P_{GS}[DS \geq S] \quad (3a)$$

$$P_{COMB}[DS \geq M] = P_{GF}[DS \geq M] + P_{GS}[DS \geq M] - P_{GF}[DS \geq M] \times P_{GS}[DS \geq M] \quad (3b)$$

$$P_{COMB}[DS \geq E] = P_{GF}[DS \geq E] + P_{GS}[DS \geq E] - P_{GF}[DS \geq E] \times P_{GS}[DS \geq E] \quad (3c)$$

$$P_{COMB}[DS \geq C] = P_{GF}[DS \geq C] + P_{GS}[DS \geq C] - P_{GF}[DS \geq C] \times P_{GS}[DS \geq C] \quad (3d)$$

where:

$P_{COMB}[DS \geq C] \leq P_{COMB}[DS \geq E] \leq P_{COMB}[DS \geq M] \leq P_{COMB}[DS \geq S] \leq 100\%$
$P_{GF}[DS \geq X]$ is the probability of having ground failure damage at least equal to damage state X, where X = Slight, Moderate, Extensive, or Complete
$P_{GS}[DS \geq X]$ is the probability of having ground shaking damage at least equal to damage state X, where X is defined as above

Example Application

The three methods for integrating the effects of ground shaking and collateral hazards are implemented in a GIS and applied to the case study region of Palo Alto, California. A scenario earthquake of a magnitude 8.0 event on the San Andreas fault at its closest location to the study region is used in the analysis. In order to simplify the comparison, several assumptions are made. The damage is estimated only for buildings of wood frame classification, damage is limited to structural damage, and only mean values are computed and compared. The regional estimates of expected damage for each hazard are first given, followed by the computation of combined or total damage distributions for the scenario earthquake event.

Strong Ground Shaking

The ATC-13 and ATC-36 methodologies do not specify a method for estimating the distribution of ground shaking in the study region. For consistency in the methodology comparison, the same ground shaking attenuation model that is specified in the NIBS methodology was implemented in the GIS and used in all three damage estimation processes. An attenuation model was used that gives PGA and spectral acceleration as a function of distance to the source, the magnitude of the earthquake, and the local soil conditions. PGA values were converted to MMI values, the required ground motion parameter for the ATC methodologies.

Ground shaking damage estimates were made for buildings of wood frame construction. The ATC-13 and ATC-36 methodologies utilize roughly the same procedure for estimating the ground shaking damage. In ATC-13 damage probability matrices giving the probability of having various damage states (with mean damage factors) for a given building class subjected to seven different levels of MMI are used. In ATC-36, the damage probability matrices are converted to expected damage factor curves and updated in some cases to account for recent empirical data, giving mean damage levels for a given building class subjected to seven different levels of MMI. Since the wood frame damage parameters are the same in ATC-13 and ATC-36 and the comparison is based on mean values, the same damage due to ground shaking is estimated using the two ATC methodologies. Figure 1 shows the expected damage due to ground shaking, EDF[S], to wood frame buildings computed according to the ATC methodologies for a magnitude 8.0 earthquake on the San Andreas fault in Palo Alto, California.

In the NIBS methodology, ground shaking damage is estimated by intersecting the demand curve (the plot of spectral acceleration versus spectral displacement) at the site with the capacity curve for a given building class (based on static nonlinear pushover analysis). The demand curve is found by implementing the attenuation model discussed above and if necessary reducing the values to account for increased damping according to the structural class. The damage to a given building class is found by using fragility curves that give the probability of being in or exceeding various damage states as a function of the intersection point found in the capacity/demand analysis. For comparison purposes, a mean damage factor can be found by multiplying the probability of each damage state by the damage factor associated with that damage state. Figure 2 shows the expected damage due to ground shaking, EDF[S], to wood frame buildings computed according to the NIBS methodology for a magnitude 8.0 earthquake on the San Andreas fault in Palo Alto.

Liquefaction

A surface geology map based on Wentworth (1993) was used to assign the liquefaction susceptibility classes to the soil deposits in the study region, a

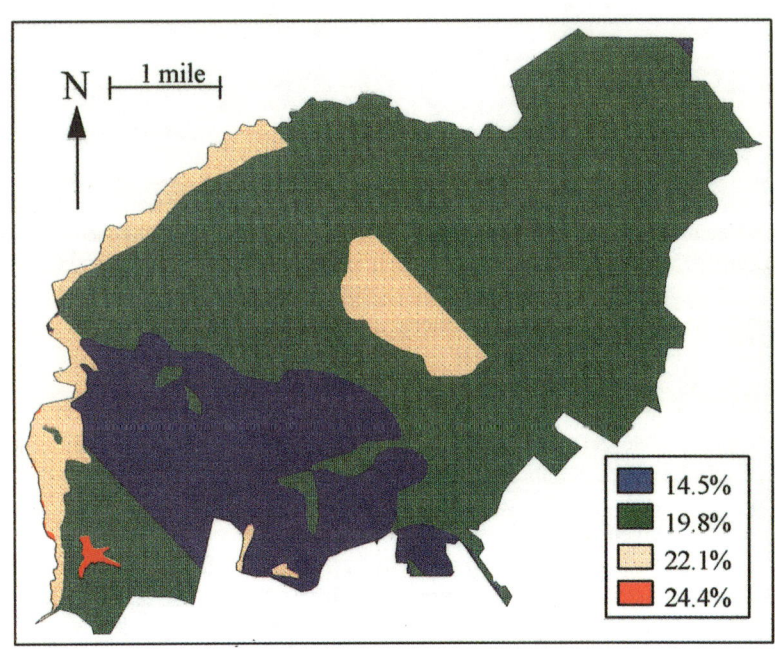

Figure 1. Wood frame building shaking damage in Palo Alto using ATC-13.

14.5%	
19.8%	
22.1%	
24.4%	

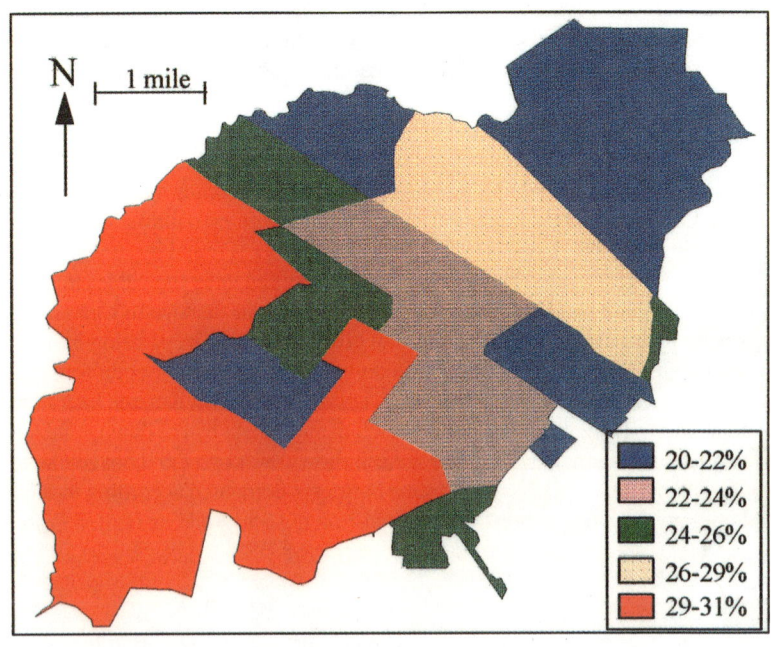

Figure 2. Wood frame building shaking damage in Palo Alto using NIBS.

20-22%	
22-24%	
24-26%	
26-29%	
29-31%	

requirement of each damage estimation methodology. Using the ATC-13 methodology, the damage due to liquefaction is a factor of five greater than the damage due to ground shaking multiplied by the probability of liquefaction (a function of the susceptibility and the input MMI level). Figure 3 shows the expected damage due to liquefaction, EDF[L], to wood frame buildings computed according to the ATC-13 methodology for a magnitude 8.0 earthquake on the San Andreas fault in Palo Alto. Using the ATC-36 methodology, the damage due to liquefaction is the damage due to ground shaking increased by one MMI level multiplied by the probability of liquefaction (a function of the susceptibility and the input MMI level). Figure 4 shows the expected damage due to liquefaction, EDF[L], to wood frame buildings computed according to the ATC-36 methodology for a magnitude 8.0 earthquake on the San Andreas fault in Palo Alto.

The permanent ground deformation due to lateral spreading, required to estimate damage due to liquefaction in the NIBS methodology, is a function of the earthquake magnitude, the input PGA, the threshold acceleration for the given susceptibility class, and the probability of liquefaction (a function of the susceptibility class, the groundwater elevation, the input PGA, and the magnitude of the earthquake). The ground water elevation for Palo Alto was estimated by averaging the values reported in boring log data that was compiled for the study region. The permanent ground deformation due to settlement is only a function of the susceptibility class. The ground deformation values are combined with those found in the landslide analysis described below.

Landslide

A slope map for the study region was constructed using DEM (Digital Elevation Model) data from the United States Geological Survey. The slope map and surface geology map were used to develop the landslide susceptibility classes that are required for landslide damage modeling in each of the three regional loss estimation methodologies. Using the ATC-13 methodology, the damage due to landslide is computed through the use of the landslide probability matrices that give the probability of various damage states as a function of input MMI level and landslide susceptibility class. Using the ATC-36 methodology, the damage due to landslide is the damage factor for the given susceptibility class (independent of building type) multiplied by the probability of landslide (a function of the susceptibility and the input MMI level). Maps of the expected damage due to landslide, EDF[LS], to wood frame buildings computed according to the ATC-13 and ATC-36 methodologies for a magnitude 8.0 earthquake on the San Andreas fault in Palo Alto are not shown here due to space limitations.

The permanent ground deformation due to landslide in the NIBS methodology is a function of the earthquake magnitude, the input PGA, the threshold acceleration for the given susceptibility class, and the probability of landslide (a function of the

111

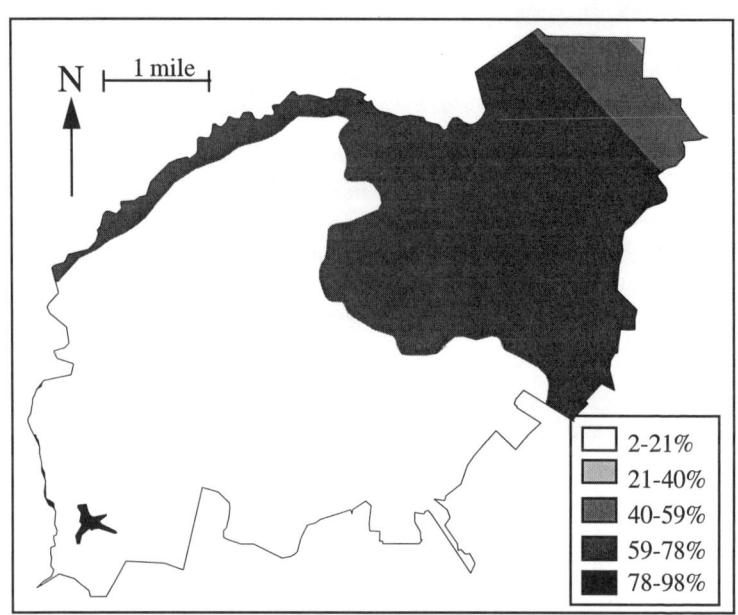

Figure 3. Wood frame building liquefaction damage in Palo Alto using ATC-13.

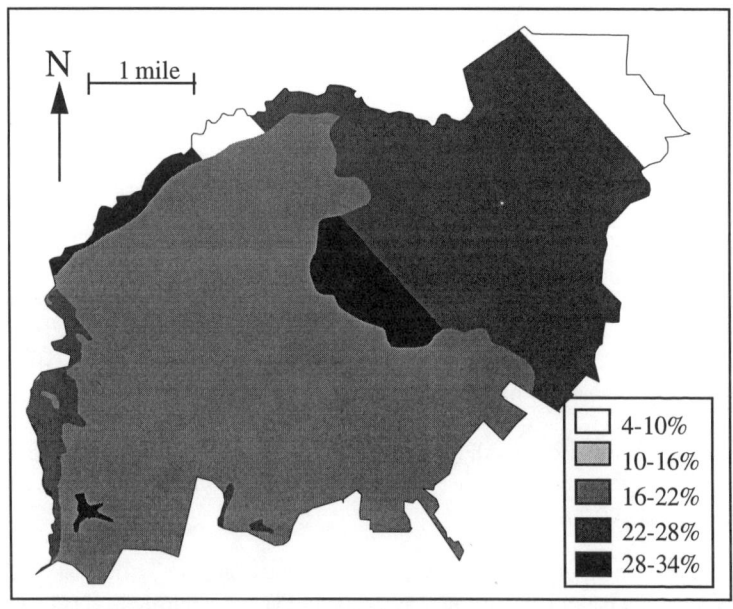

Figure 4. Wood frame building liquefaction damage in Palo Alto using ATC-36.

susceptibility class). The permanent ground deformation caused by landslide is added to that due to lateral spreading caused by liquefaction. Using a fragility curve that is independent of building type, damage due to ground failure is estimated as the probability of having at least extensive building damage based on the amount of lateral spreading and settlement. Figure 5 shows the probability of having at least extensive damage due to landslide according to the NIBS methodology for a magnitude 8.0 earthquake on the San Andreas fault in Palo Alto.

Integrated Damage Estimates

Using the ATC-13 damage combination methodology (Equation 1), the total damage that would be expected for a wood frame building in Palo Alto due to a magnitude 8.0 earthquake on the San Andreas fault is as shown in Figure 6. The total expected damage ranges from 30% to 100% using the ATC-13 method. In some cases, the total expected damage had to be truncated to 100%. Figure 7 shows the total damage that would be expected for a wood frame building in Palo Alto due to a magnitude 8.0 earthquake on the San Andreas fault computed according to the ATC-36 method (Equation 2). The total damage ranges from 22% to 77% in Figure 7. Using the NIBS damage combination methodology (Equation 3), the total damage that would be expected for a wood frame building in Palo Alto due to a magnitude 8.0 earthquake on the San Andreas fault is as shown in Figure 8. In this case, the total damage ranges from 12% to 36%.

Conclusions

A comparison of Figures 1 and 2 shows that the wood frame building shaking damage is slightly higher with the NIBS methodology than with the ATC-13 and ATC-36 methodologies. The mean values are fairly close and certainly within the range of uncertainty associated with the estimates. A comparison of Figures 3 and 4 shows that liquefaction damage to wood frame buildings computed with the ATC-13 method is significantly higher that that computed with the ATC-36 method. In addition, the ATC-13 liquefaction damage is more concentrated in the area of softer soils, while the ATC-36 liquefaction damage is more evenly distributed throughout the study region. Similar to the liquefaction damage estimated with the ATC-13 method, the NIBS permanent ground deformation damage shown in Figure 5 is fairly concentrated in the area of softer soils and is relatively high only in that region.

Figures 6, 7, and 8 illustrate that the ATC-13 damage combination method produces the highest damage estimates, as is expected because of the straight sum approach to the damage integration. The ATC-36 damage combination method produces total damage estimates that are lower than those from the ATC-13 analysis, but are significantly higher than the total damage estimates computed with the NIBS method. In addition, the ATC-36 damage estimates appear to be heavily influenced by the surface geology map (not shown in this paper), as the areas of the damage

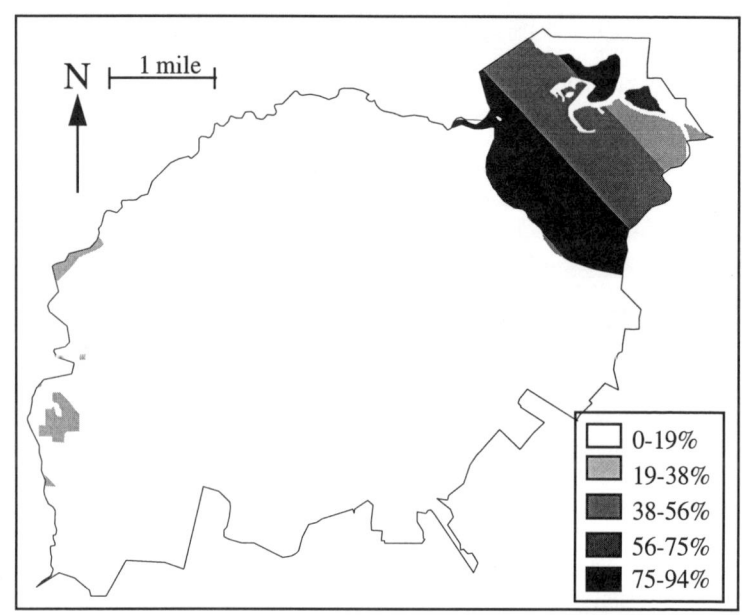

Figure 5. Probability of at least extensive ground failure damage in Palo Alto.

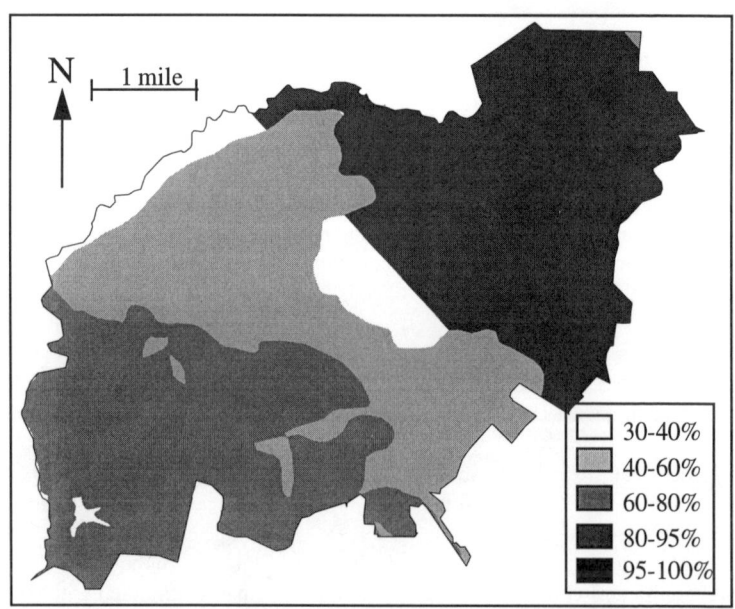

Figure 6. Wood frame building combined damage in Palo Alto using ATC-13.

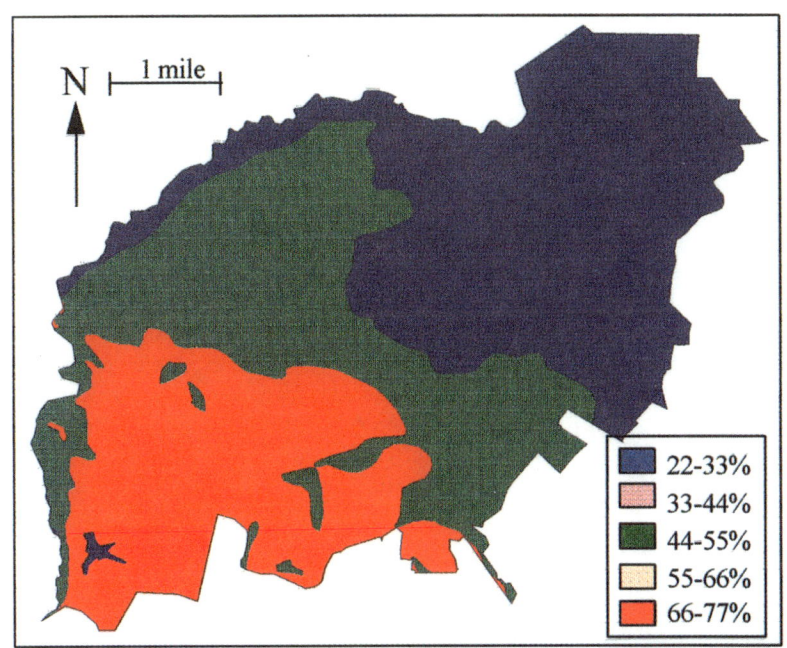

Figure 7. Wood frame building combined damage in Palo Alto using ATC-36.

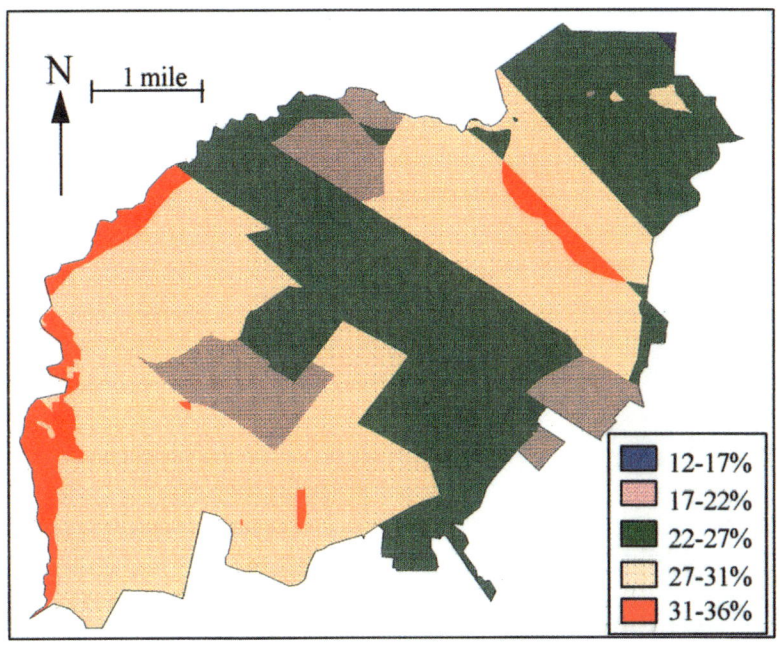

Figure 8. Wood frame building combined damage in Palo Alto using NIBS.

regions and different geologic units are almost exactly the same. This similarity requires further analysis. The NIBS combination methodology produces the lowest total damage results. With the NIBS method, the total damage estimates are only 5% higher than the shaking damage estimates, indicating that the ground failure damage has relatively little effect on the average damage factor, except in areas of poor soils.

The results indicate that the choice of earthquake damage combination method can have a significant impact on the total damage and loss estimates in a region, even though the damage due to ground shaking is nearly the same using the different regional loss estimation methodologies. These results are only applicable to mean estimates of wood frame damage in Palo Alto for a magnitude 8.0 event on the San Andreas fault. Further analysis for different building types in Palo Alto, as well as other seismic regions, might provide different results. In addition, the effects of other scenario events and probabilistic hazard analyses should be considered.

Acknowledgments

The work presented in this paper was funded by the Kajima Corporation through the California Universities for Research in Earthquake Engineering (CUREe) and the John A. Blume Earthquake Engineering Center at Stanford University. In addition, ESRI provided their Arc/Info GIS software for use in the analysis. The support of these organizations is appreciated.

References

Applied Technology Council (1985). *Earthquake Damage Evaluation Data for California*, ATC-13 report, Redwood City, CA.

Applied Technology Council (in progress). *Earthquake Loss Evaluation Methodology and Databases for Utah*, ATC-36 report, Redwood City, CA.

King, S.A. and Kiremidjian, A.S. (1994). *Regional Seismic Hazard and Risk Analysis Through Geographic Information Systems.* Technical Report No. 111. John A. Blume Earthquake Engineering Center, Stanford University, Stanford, CA.

Kiremidjian, A.S. (1992). Methods for Regional Damage Estimation. *Special Proceedings of the 10th World Conference on Earthquake Engineering*, Madrid.

Risk Management Solutions (1985). *Development of a Standardized Earthquake Loss Estimation Methodology*, prepared forNIBS, Menlo Park, CA.

Wentworth, C.M. (1993). *General Distribution of Geologic Materials in the Southern San Francisco Bay Region, California: A Digital Database*, Open-File Report 93-393. United States Geological Survey, Menlo Park, CA.

GIS ASSESSMENT OF WATER SUPPLY DAMAGE
FROM THE NORTHRIDGE EARTHQUAKE

by

T.D. O'Rourke[1]

S. Toprak[2]

ABSTRACT

The 1994 Northridge earthquake resulted in the most extensive damage to a US water supply system since the 1906 San Francisco earthquake. Three major transmission systems, which provide over three-quarters of the water for the City of Los Angeles, were disrupted. Los Angeles Department of Water and Power (LADWP) and Metropolitan Water District (MWD) trunk lines (nominal pipe diameter $\geq$ 600 mm) were damaged at 74 locations, and the LADWP distribution system required repairs at 1013 locations. The widespread disruption provides a unique opportunity to evaluate the geographic variability of the damage, the most vulnerable pipelines, and the relationship among damage, transient motion, and permanent ground deformation.

This paper describes how a geographical information system (GIS) was developed for the earthquake response of the LADWP system. Damage patterns are evaluated for each type of pipeline and related to seismic intensity, measured transient motion, locations of high water table, and locations of permanent ground deformation. Key factors affecting earthquake damage are discussed.

INTRODUCTION

The 1994 Northridge earthquake led to significant disruption of the water supply system of Los Angeles, causing damage at 15 locations in the three transmission systems providing water from Northern California, 74 locations in water trunk lines (nominal pipe diameter $\geq$ 600 mm), and 1013 locations in the Los

---

1, 2 - Professor and Graduate Research Assistant, respectively, Cornell University, School of Civil & Env. Eng., Hollister Hall, Ithaca, NY 14853-3501

Angeles Department of Water and Power (LADWP) distribution pipeline network. The damage was distributed over approximately 1200 km^2. Damage of such a widespread nature invites questions about its spatial variability and its relationship with parameters such as earthquake intensity, peak acceleration, peak velocity, groundwater levels, and areas of permanent ground deformation (PGD). Large water supply systems are composed of pipelines constructed with different materials, diameters, and joint characteristics. Hence, there are questions about how the damage patterns are influenced by different material and mechanical characteristics.

Questions related to spatial variability are well suited for evaluation with geographical information systems (GIS). This paper provides a description of how a GIS database was assembled for water supply damage caused by the Northridge earthquake. The characteristics of the Los Angeles water supply are discussed, and graphical information is provided regarding the overall statistics and spatial patterns of pipeline damage. Various spatial relationships between earthquake damage and seismic intensity are explored. Local patterns of damage and repair also are examined relative to groundwater levels and zones of liquefaction-induced PGD.

EARTHQUAKE DAMAGE DATABASES

The earthquake-induced damage to water pipelines and the database developed to characterize this damage have been described by O'Rourke, et al. (1996), and only the salient features of this work are summarized here. Geographic information system databases for lengths of trunk lines according to pipe composition and size were assembled with ARC/INFO software (ESRI, Inc., 1994). All LADWP and Metropolitan Water District (MWD) trunk lines within the area covered by the LADWP system were digitized from 1:12,000 scale pipeline maps provided by LADWP.

The trunk line repair database was assembled from repair statistics provided by LADWP and MWD, as well as extensive discussions with engineers of both organizations. The distribution line repair database was organized from repair statistics developed for the State of California Office of Emergency Services (OES). After a careful evaluation of the 1405 original OES repair records, it was determined that 1013 were valid for damage at distribution mains and hydrants. Reliable information about pipe composition could be found for 964 repairs, and this portion of the database is used in this paper for evaluating percentages of pipeline damage according to composition.

The distribution pipeline database was supplemented by databases developed for the trunk line and aqueduct systems. Repairs to all LADWP and MWD trunk lines within the City of Los Angeles were characterized and located in the GIS. The repairs to all transmission pipelines upstream of the Jensen and City of Los Angeles water treatment facilities were similarly assembled in the GIS database.

LOS ANGELES WATER SUPPLY SYSTEM

Figure 1 shows that the portion of the Los Angeles water supply system most seriously affected by the Northridge earthquake superimposed on the topography of Los Angeles. The water supply system includes transmission lines, trunk lines and distribution lines. All large diameter pipelines upstream of the treatment plants are considered to be transmission facilities.

Figure 2 presents charts that show the relative lengths of LADWP and MWD trunk and distribution lines according to pipe composition. It should be noted that the vertical axis in Figure 2c is logarithmic scale. The figures were developed from LADWP statistics, valid as of June, 1994. Because statistics on ductile iron (DI) pipe are not provided in LADWP summaries of total pipeline length, the percentage of DI pipe was estimated on the basis of evaluations of the digitized portion of the system. The MWD trunk lines include pipelines within the area of the LADWP system designated by MWD as feeder lines. Total lengths of approximately 700 km and 300 km for LADWP and MWD trunk lines, respectively, are included in the database, and the pie charts in each figure show the relative percentages of the combined LADWP and MWD trunk lines associated with different types of pipe material. The concrete category includes precast concrete, prestressed concrete, reinforced concrete, and concrete cylinder pipelines. About 70% of the trunk lines are composed of steel, with 14% associated with riveted steel and 56% associated primarily with steel pipe connected by welded slip joints. The total length of the distribution lines is 10,750 km. About 76% and 11% of the distribution lines composed of cast iron and steel, respectively.

Figure 3 shows cast iron (CI) pipeline density in the study area. Pipeline density is presented in terms of pipe length contours, obtained by overlaying a 2 km x 2 km mesh onto distribution lines and determining the CI pipeline length in each mesh area. The length of pipelines in each area is divided by the same area to provide a normalized parameter independent of grid size. Contours then were drawn from the spatial distribution of normalized lengths, each of which were centered on its tributary area. Since 76% of the water distribution system consists of CI pipelines with widespread distribution throughout the city, this type of pipeline can be used most effectively to evaluate the distribution and characteristics of damage over the entire area.

Figure 4 shows pipe length contours for steel distribution pipelines, developed in a manner similar to that for Figure 3. Comparison of Figures 3 and 4 shows a significant difference between the density patterns for CI and steel pipe. Steel pipelines are concentrated mostly in hillsides and mountains.

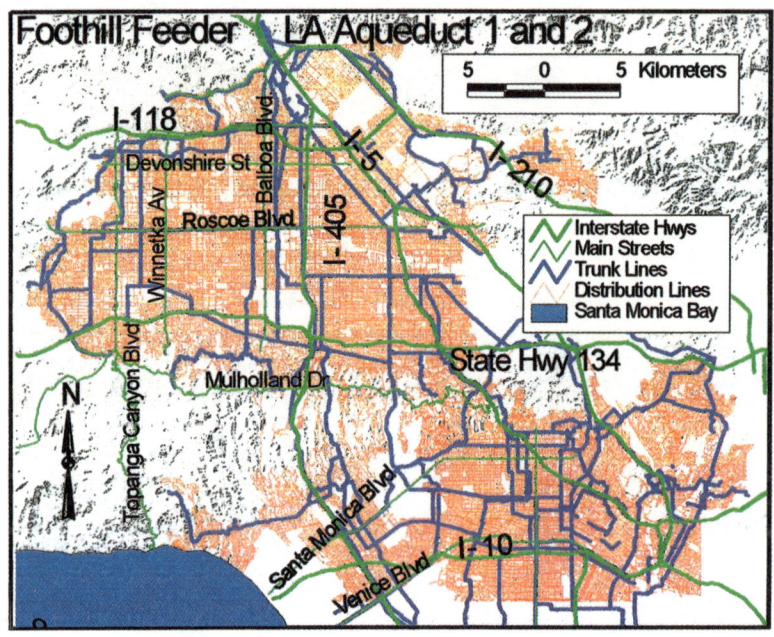

Figure 1. Map of Los Angeles Water Supply System Affected by Northridge Earthquake

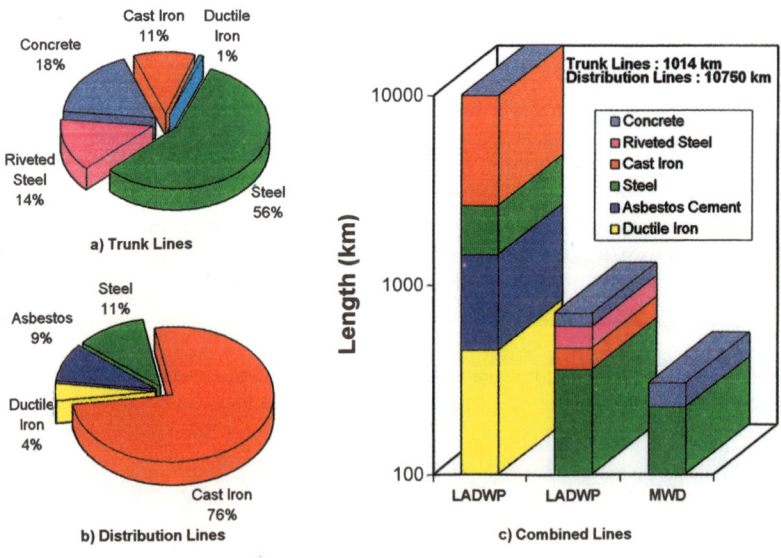

Figure 2. Composition Statistics of Water Trunk and Distribution Lines

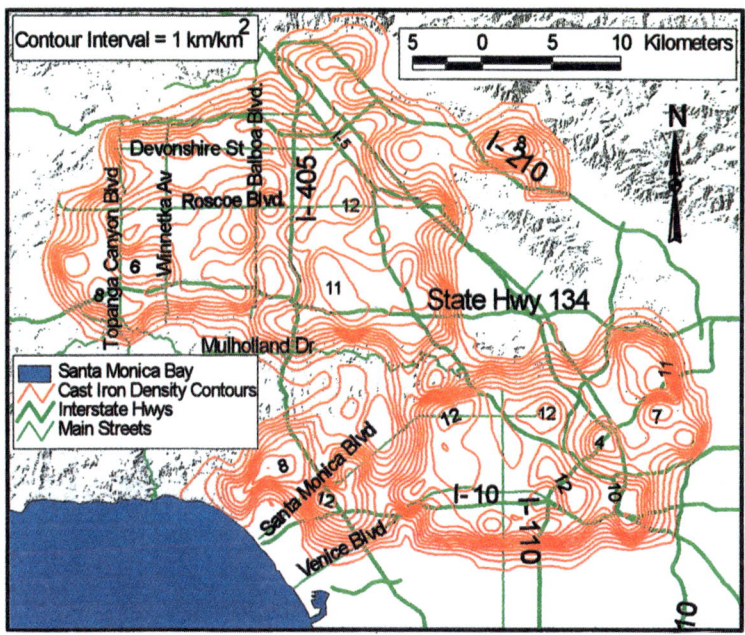

Figure 3. Cast Iron Pipeline Density Map

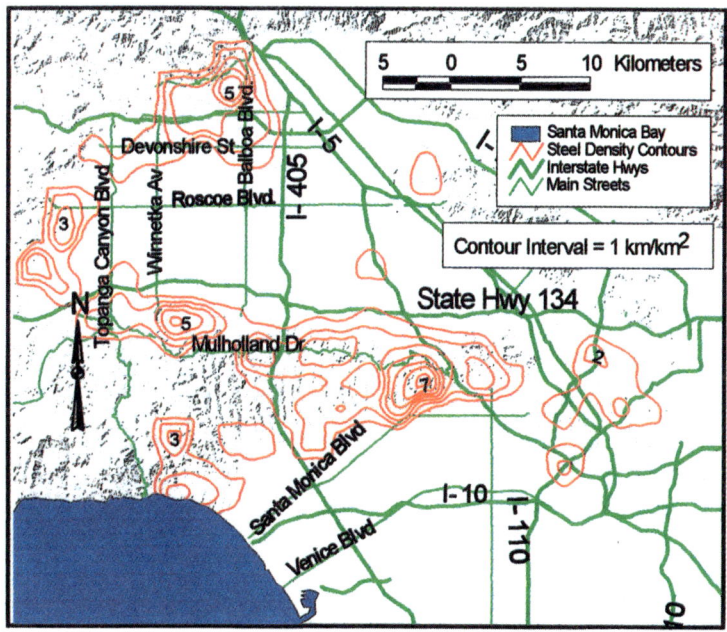

Figure 4. Steel Pipeline Density Map

NORTHRIDGE EARTHQUAKE DAMAGE

Figure 5 shows the locations of transmission lines and associated repairs superimposed on a topographic map. The map illustrates an area to the north of the City of Los Angeles. Elevation contours at 50 m intervals were obtained from Digital Elevation Model (DEM) files. Locations of repairs are marked by solid circles, with the numbers of repairs indicated at locations of concentrated damage. Locations of welded slip joint repairs are identified separately. As shown in the figure, the aqueduct systems that supply water from northern California are the Foothill Feeder, operated by MWD, and Los Angeles Aqueducts No. 1 and 2, operated by LADWP.

A description of damage sustained by LADWP and MWD transmission lines has been provided by Lund (1995). In total, there were 11 repair locations in LADWP Aqueducts No. 1 and 2, of which 4 involved either circumferential cracks or compressive buckling at welded slip joints. Excessive axial slip occurred at 2 Dresser couplings. Damage to the MWD transmission system occurred near the Jensen Filtration Plant at a welded slip joint of a 2160-mm steel pipeline and as cracks and leakage in a reinforced concrete conduit. Damage was sustained by another steel pipeline at a sleeve-type coupling and in an area of differential settlement and horizontal movement adjacent to the Jensen Plant.

Figure 6 presents a map of trunk line repair locations in the northwestern portion of San Fernando Valley. Sixty-seven of the 74 trunk line repairs were located in this region, with the highest concentrations of damage in the Van Norman Complex, near the intersection of Balboa Blvd. and Rinaldi St., and along Roscoe Blvd. At locations of multiple repairs, a number is shown that represents the sum of repairs made at that location.

Figure 7 presents a map of distribution pipeline repair locations and repair rate contours for CI pipeline damage. The repair rate contours were developed by dividing the map into 2 km x 2 km areas, determining the number of CI pipeline repairs in each area, and dividing the repairs by the distance of CI main in that area. Contours then were drawn from the spatial distribution of repair rates, each of which was centered on its tributary area. The 2 km x 2 km grid was found to provide a good representation of damage patterns for the map scale of the figure. These contours are especially well suited for comparing damage with the spatial distribution of strong motion parameters, as discussed in a forthcoming section.

Figure 8 presents a map of pipeline repair locations and repair rate contours for steel pipelines. The map and contours were developed in a manner similar to that for the CI pipeline repair rate contours. Both Figures 7 and 8 need to be evaluated relative to the spatial patterns of density for CI and steel pipelines presented in Figures 3 and 4, respectively. For example, Figure 4 shows that steel

122

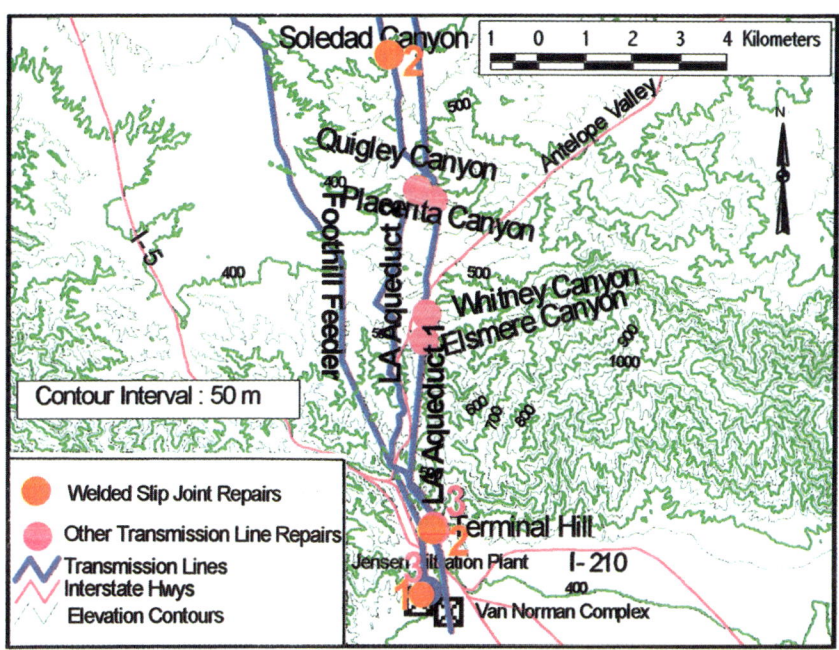

Figure 5. Water Transmission Line Damage

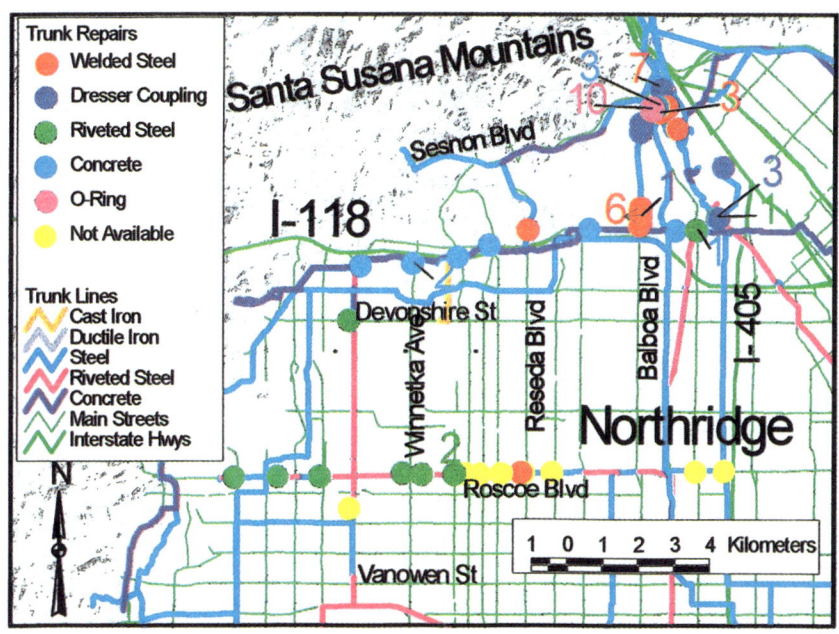

Figure 6. Water Trunk Line Damage

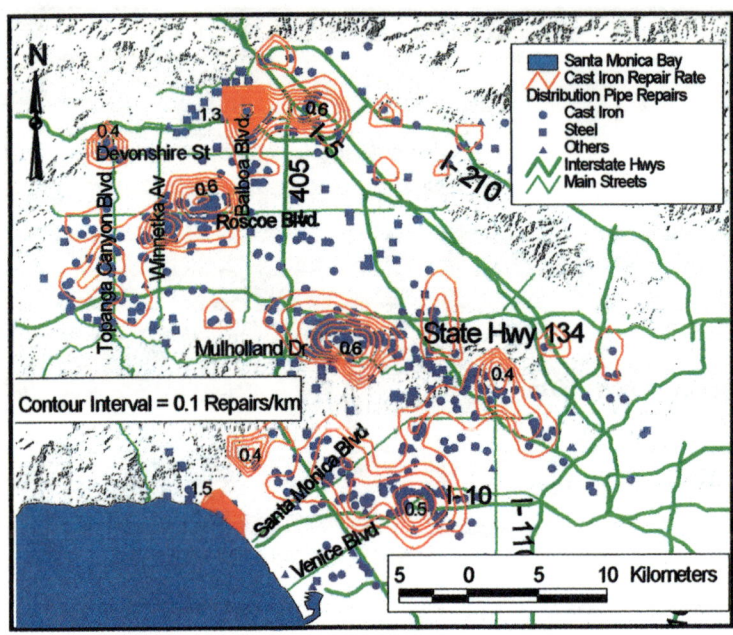

Figure 7. Cast Iron Pipeline Repair Rate Contours

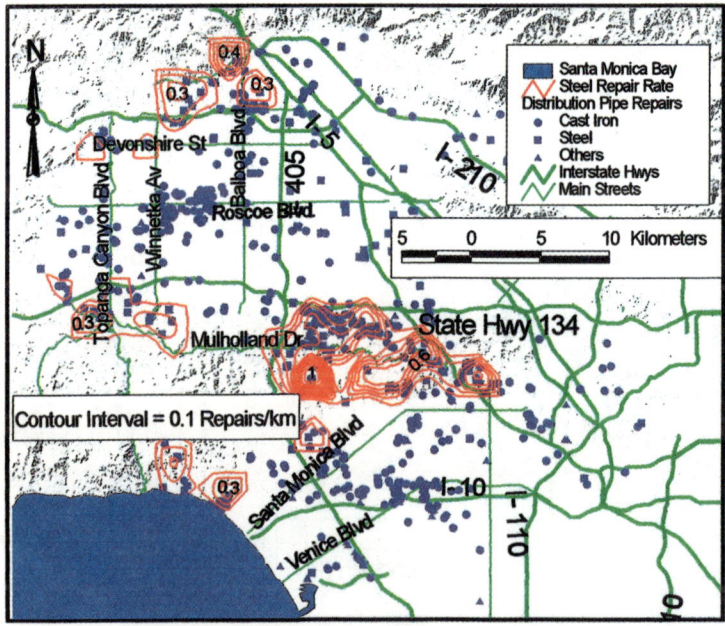

Figure 8. Steel Pipeline Repair Rate Contours

pipelines are located mostly at higher elevations surrounding the San Fernando Valley. Steel pipeline repairs, therefore, are located in these areas.

The overall statistics of pipeline damage are summarized in Figure 9 in the form of pie and bar charts. As with Figure 2, please note that the bar chart scale is logarithmic. Most repairs to trunk lines occurred in steel pipelines, with 80% of all repairs in riveted and continuous wall steel piping. Sixty-six percent of repairs were in continuous wall steel pipe, whereas only 56% of all trunk lines were composed of this type of pipe. The steel trunk lines were damaged heavily by compressive wrinkling of welded slip joints, with this type of damage recorded at 20 locations. There also were 10 reported locations of pullout at compression couplings.

Seventy-one percent of the distribution line repairs were in CI pipelines, which constitute 76% of the system (see Figure 2). Twenty-two percent of the repairs were in steel pipelines, which constitute only 11% of the system (see Figure 2). The relatively high concentration of steel pipeline repairs is associated with various types of steel, such as Mannesman and Matheson steel, which are prone to corrosion, as well as damage at certain types of elastomeric joints that are vulnerable to creep and leakage.

EARTHQUAKE DAMAGE VS. SEISMIC INTENSITY

Figure 10 shows the repair rate contours for CI pipelines superimposed on zones of Modified Mercalli Intensity (MMI) mapped by USGS (Dewey, et al., 1995). The locations of highest repair rate coincide with areas of MMIX and MMVIII.

Figures 11 and 12 show the repair rate contours for CI pipelines superimposed on zones of peak acceleration and velocity measured by free-field strong motion instruments. The free-field records used to plot peak acceleration and velocity zones are identical to those described by Chang, et al. (1996) in their evaluation of the engineering implications of the earthquake motion. The records from approximately 240 rock and soil stations were used. The maximum horizontal acceleration of 1.78 g measured at the Tarzana-Cedar Hill Nursery was removed from the database prior to GIS evaluation to avoid distortions from possible topographic influences. In addition, records from stations at dam abutments were screened when a station downstream of the dam was available, again to minimize distortion from topographic effects.

The zones of highest peak acceleration coincide reasonably well with the locations of highest repair rate, especially near the northern edge of the San Fernando Valley, the Santa Monica Mountains, and the Los Angeles Basin. The zones of highest peak velocity show similar spatial correlation with repair rate concentrations. The velocities do not coincide as well with repair concentrations in

125

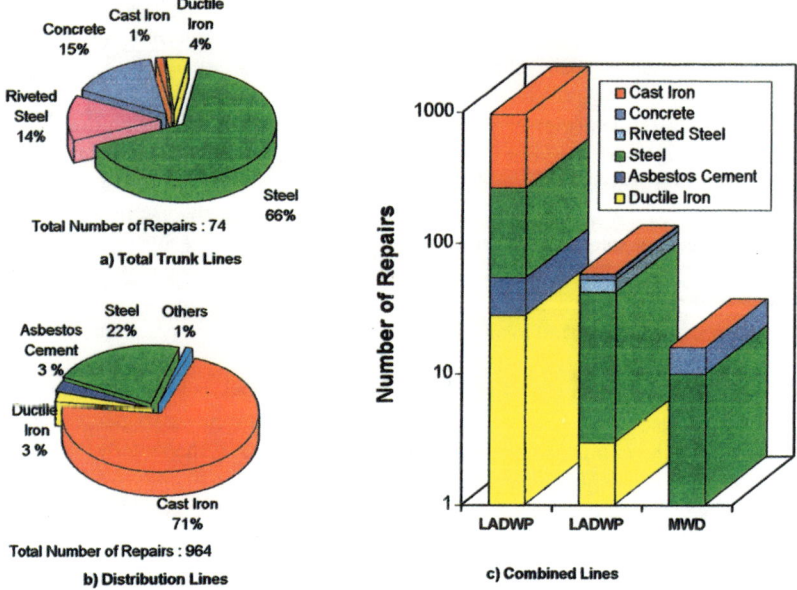

Figure 9. Repair Statistics of Water Trunk and Distribution Lines

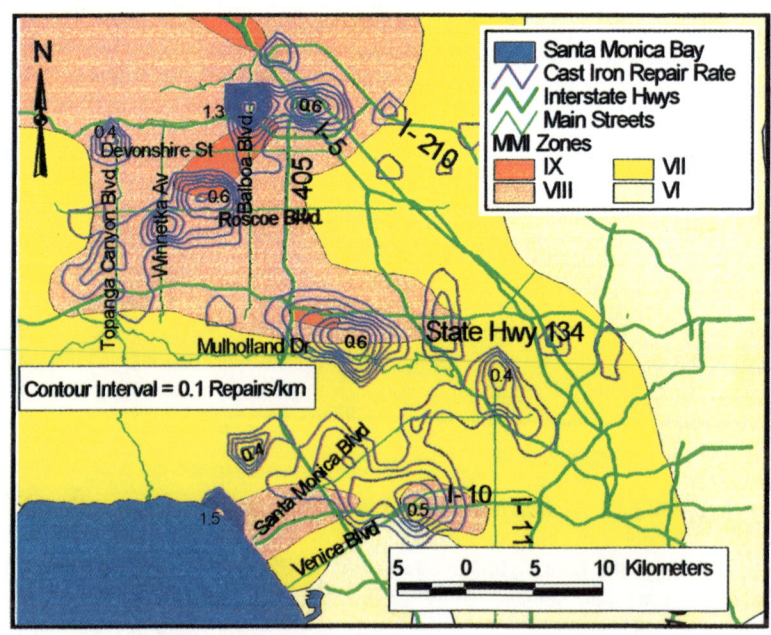

Figure 10. Pipeline Repair Rate Contours vs. MMI

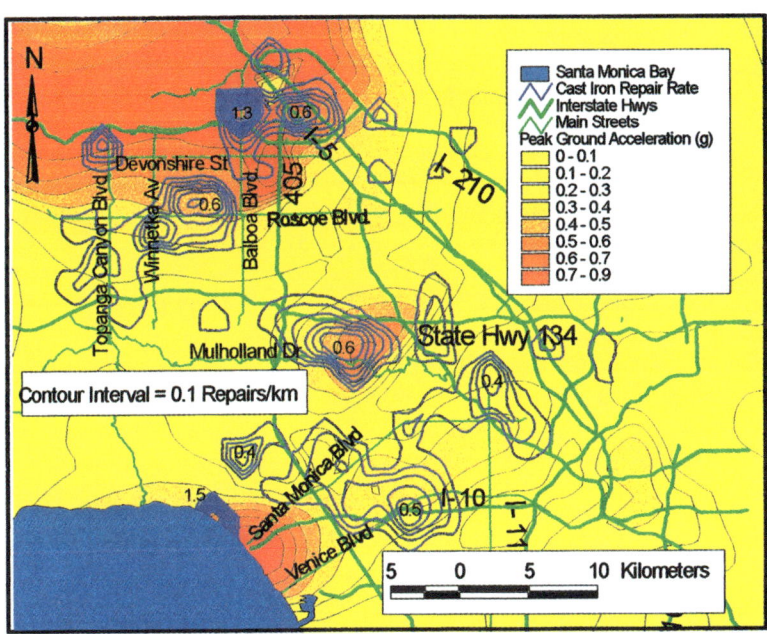

Figure 11. Pipeline Repair Rate Contours vs. Peak Ground Acceleration

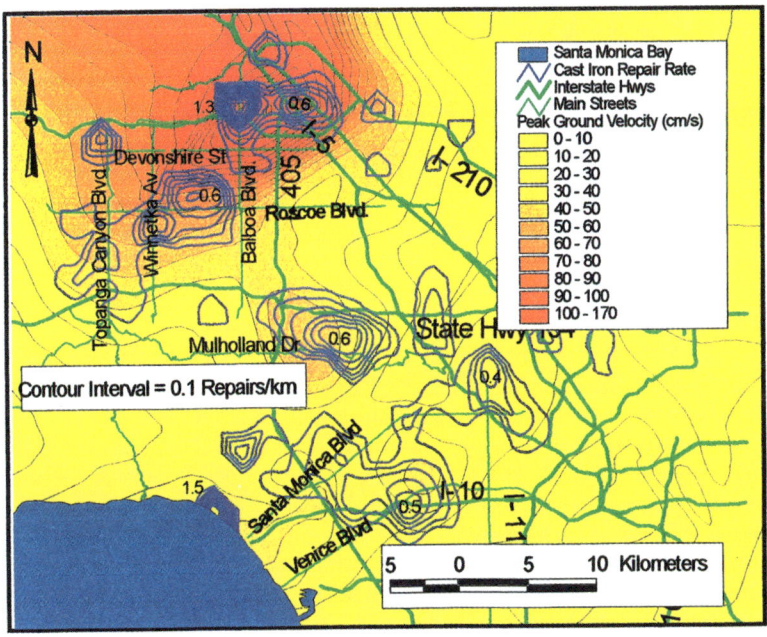

Figure 12. Pipeline Repair Rate Contours vs. Peak Ground Velocity

the Los Angeles Basin, but correlate better with CI pipeline repairs in the western and central portions of San Fernando Valley.

It is notable that high concentrations of repair rate occur in the northern portion of the Santa Monica Mountains in the Sherman Oaks area. This location coincides with a zone of extensive slope movements, ground fissures, and cracking of artificial fill described by Barrows, et al. (1995). Ground failure of this type is likely affected by acceleration levels, which were very high in the Sherman Oaks area, as illustrated in Figure 11. Zones of high acceleration, therefore, would be expected to correlate with locations of ground failure and thus PGD effects on pipelines. In contrast, zones of high velocity would be expected to correlate well with locations of high transient ground strain.

LOCAL PATTERNS OF DAMAGE

Figure 13 shows CI pipeline repair rate contours in the western and central portions of San Fernando Valley superimposed on the outline of high groundwater level zones that are taken from Tinsley, et al. (1985) and Los Angeles County (1990). The highest repair rate concentrations occur in pre- and post-1944 zones with groundwater approximately 3 m deep. Holzer, et al. (1996) report that locations within this area experienced PGD from both liquefaction and failure of soft clay sediments. It appears that the high groundwater table helps to delineate locations of liquefiable sands and soft clay, both of which are susceptible to ground failure as well as large transient strains associated with site amplification.

Figure 14 shows locations of pipeline repair and repair rate contours near the intersection of Balboa Blvd. and Rinaldi St. This area was influenced by liquefaction-induced lateral spread (Holzer, et al., 1996) that contributed to catastrophic failure of a gas transmission and two water trunk lines (O'Rourke and Toprak, 1995). A 0.5 km x 0.5 km was used to determine repair rates in this area. The local repair rates are exceptionally high, with a maximum value of 4.5 repairs/km.

CONCLUSIONS

Geographical information systems (GIS) are well suited for evaluating the spatial relationships between water supply system damage and factors such as seismic intensity, peak acceleration, peak velocity, groundwater levels, and locations of ground failure. Spatial correlations between water pipeline damage and each of these parameters were examined for the 1994 Northridge earthquake. Good correlations between pipeline repair rates and both peak acceleration and peak velocity were found, although neither parameter was observed to provide consistently strong correlations at all locations of concentrated repair. It appears that zones of high peak acceleration coincide with locations of ground failure and thus PGD effects on pipelines. In contrast, zones of high velocity would be

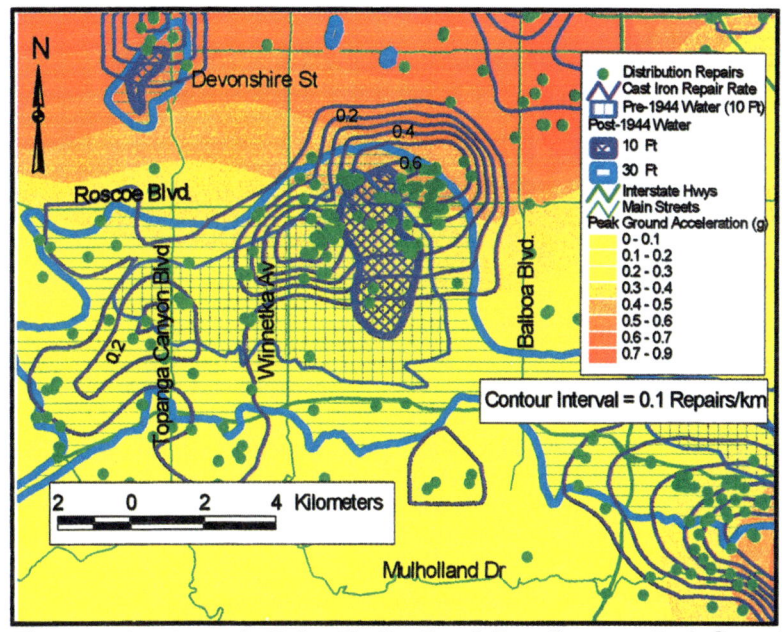

Figure 13. San Fernando Valley Pipeline Repair Rate Contours, Peak Ground Acceleration, and Ground Water Table

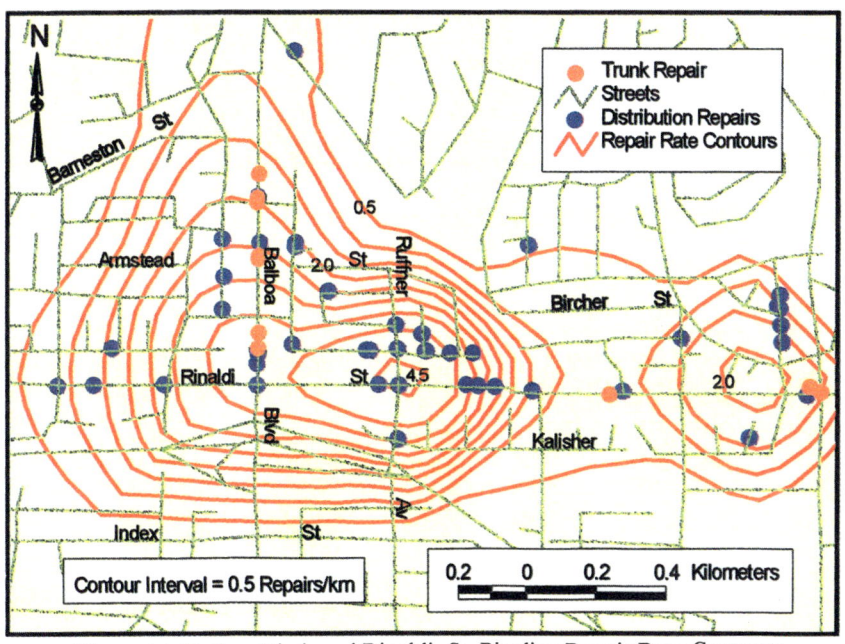

Figure 14. Balboa Blvd. and Rinaldi St. Pipeline Repair Rate Contours

expected to correlate well with locations of high transient ground strains. Zones of high groundwater level in the western and central portions of San Fernando Valley help to delineate locations of liquefiable sands and soft clay, both of which are susceptible to ground failure as well as large transient strains associated with site amplification. There is a strong spatial correlation between pipeline repair rates resulting from the Northridge earthquake and high groundwater levels in the San Fernando Valley.

ACKNOWLEDGMENTS

The research reported in this paper was supported by the National Center for Earthquake Engineering Research, Buffalo, NY under Contract Nos. R91598 and R91605. The authors wish to express their deep gratitude to Mr. H. Dekermenjian of LADWP for his interest and assistance in obtaining pipeline and pipeline repair data. Thanks are also extended to Mr. N. Blaze of EQE, Inc., Mr. C. Davis of LADWP, and Mr. D. Wright of MWD for their assistance in obtaining valuable information. The manuscript was prepared by L. McCall, and A. Avcisoy helped prepare the figures.

REFERENCES

Barrows, A.G., Irvine, P.J., and Tan, S.S., "Geologic Surface Effects Triggered by the Northridge Earthquake," The Northridge, California, Earthquake of 17 January 1994, Special Publication 116, M.C. Woods and W.R. Seiple, Eds., California Division of Mines and Geology, Sacramento, CA, 1995, pp. 65-88.

Chang, S.W., Bray, J.D., and Seed, R.B., "Engineering Implications of Ground Motions from the Northridge Earthquake," Bulletin of the Seismological Society of America, Vol. 86, No. 1B, El Cerrito, CA, Feb. 1996, pp. 5270-5288.

Dewey, J.W., Reagor, B.G., Dengler, L., and Moley, K., "Intensity Distribution and Isoseismal Maps for the Northridge, California, Earthquake of January 17, 1994," U.S. Geological Survey Open-File Report 95-92, U.S. Department of the Interior, Washington, DC, 1995.

ESRI, The ARC/INFO User's Guide, Environmental Systems Research Institute, Inc., Redlands, CA, 1994.

Holzer, T.L., Bennett, M.J., Tinsley, J.C. III, Ponti, D.J., and Sharp, R.V., "Causes of Ground Failure in Alluvium during the Northridge, California Earthquake of January 17, 1994," Technical Report NCEER-96-0012, National Center for Earthquake Engineering Research, Buffalo, NY, 1996, pp. 345-360.

Los Angeles County, Technical Appendix to the Safety Element of the Los Angeles County General Plan, Hazard Reduction in Los Angeles County, Department of Regional Planning, Los Angeles, CA, Dec. 1990.

Lund, L., "Water Systems," in Northridge Earthquake Lifeline Performance and Post-Earthquake Response, A.J. Schiff, Ed., TCLEE Monograph No. 8, ASCE, New York, NY, August 1995, pp. 96-131.

O'Rourke, T.D. and Toprak, S., "Case History of Pipeline Response to Ground Deformation at Balboa Blvd., 1994 Northridge Earthquake," Proceedings, Sixth U.S.-Japan Workshop on Earthquake Disaster Prevention for Lifeline Systems, Osaka, Japan, July 1995, pp. 3-20.

O'Rourke, T.D., Toprak, S., and Sano, Y., "Los Angeles Water Pipeline System Response to the 1994 Northridge Earthquake," Technical Report NCEER-96-0012, National Center for Earthquake Engineering Research, Buffalo, NY, 1996, pp. 1-16.

Tinsley, J.C., Youd, T.L., Perkins, D.M., and Chen, T.F., "Evaluation of Liquefaction Potential," Evaluating Earthquake Hazards in the Los Angeles Region - An Earth-Science Perspective, U.S. Geological Survey Professional Paper 1360, J.I. Ziony, Ed., U.S. Department of the Interior, Washington, DC, 1985, pp. 263-316.

SUBJECT INDEX

Page number refers to the first page of paper

NEW GEOTECH BOOKS FROM GEOLOGAN
JULY 15-19, 1997 ● LOGAN, UTAH

Dredging and Management of Dredged Material
● Jay N. Meegoda, Thomas H. Wakeman III, Arul K. Arulmoli, and William J. Librizzi, Editors
Geotechnical Special Publication #65
Proceedings of three sessions held in conjunction with Geo-Logan sponsored by the Soil Properties Committee of The Geo-Institute of the ASCE
208 pp, List, $25.00; ASCE member, **$18.75** (#40254)

Observation and Modeling in Numerical Analysis and Model Tests in Dynamic Soil-Structure Interaction
● Toyoaki Nogami, Editor
Geotechnical Special Publication #64
Proceedings of sessions held in conjunction with Geo-Logan sponsored by The Geo-Institute of the ASCE
152 pp, List, $22.00; ASCE member, **$16.50** (#40252)

Ground Improvement, Ground Reinforcement, Ground Treatment: Developments 1987-1997
● Vernon R. Schaefer, Editor
Geotechnical Special Publication #69
Proceedings of sessions sponsored by the Committee on Soil Improvement and Geosynthetics of The Geo-Institute of the ASCE in conjunction with Geo-Logan '97
632 pp, List, $53.00; ASCE member, **$39.75** (#40260)

Spatial Analysis in Soil Dynamics and Earthquake Engineering
● J. David Frost, Editor
Geotechnical Special Publication #67
Proceedings of sessions held in conjunction with Geo-Logan '97 sponsored by The Geo-Institute of the ASCE
144 pp, List, $40.00; ASCE member, **$30.00** (#40258)

Grouting: Compaction, Remediation and Testing
● C. Vipulanandan, Editor
Geotechnical Special Publication #66
Proceedings of sessions sponsored by the Grouting Committee of The Geo-Institute of the ASCE in conjunction with the GeoLogan 97 Conference
352 pp, List, $35.00; ASCE member, **$26.25** (#40255)

Unsaturated Soil Engineering Practice
● Sandra L. Houston and Delwyn G. Fredlund, Editors
Geotechnical Special Publication #68
Committee Report by the Subcommittee on Unsaturated Soils and the Committee on Shallow Foundations of The Geo-Institute of the ASCE; Proceedings of sessions on Unsaturated Soils
344 pp, List, $34.00; ASCE member, **$25.50** (#40259)

ASCE
American Society of Civil Engineers

American Society of Civil Engineers 1801 Alexander Bell Drive, Reston, VA 20191-4400
Phone 800.548.2723 (ASCE), 703.295.6300; Fax 703.295.6333; email marketing@asce.org

AUTHOR INDEX

Page number refers to the first page of paper